BIOCHEMISTRY RESEARCH TRENDS

BIOCHEMISTRY RESEARCH SUMMARIES (WITH BIOGRAPHICAL SKETCHES)

VOLUME 1

Biochemistry Research Trends

Additional books in this series can be found on Nova's website
under the Series tab.

Additional e-books in this series can be found on Nova's website
under the eBooks tab.

BIOCHEMISTRY RESEARCH TRENDS

BIOCHEMISTRY RESEARCH SUMMARIES (WITH BIOGRAPHICAL SKETCHES)

VOLUME 1

JAMES T. WALKER
AND
CATHERINE M. JONES
EDITORS

nova publishers
New York

Library of Congress Cataloging-in-Publication Data

ISBN: 978-1-53610-895-8

Published by Nova Science Publishers, Inc. † New York

Contents

PREFACE

This new book compiles biographical sketches of top professionals in the field of biochemistry, as well as research summaries from a number of different focuses in this important field.

PART I

RESEARCHER BIOGRAPHICAL SKETCHES

In: Biochemistry Research Summaries. Volume 1
Editors: James T. Walker and Catherine M. Jones

ISBN: 978-1-53610-895-8

Chapter 1

KHALED M. ABASS

Affiliation:
University of Oulu, Finland

Date of Birth:
03/1975

Education:
PhD, European Registered Toxicologist

Research and Professional Experience:

He received his PhD from Institute of Biomedicine, University of Oulu, Finland (2005-2010), specializing in environmental toxicology and human health risk assessment. His publications are focused on characterizations of human cytochrome P450 system interactions with environmental contaminants and the integrations of in vitro hepatic models data into human risk assessment – a topic he has been working on for 10 years. In 2011, he has accorded the title of a Certified European Registered Toxicologist 'ERT' by the Federation of European Toxicologists & European Societies of Toxicology EUROTOX. Recently his main field of research is focusing on environmental health and toxicology, toxicokinetics and human health risk assessment.

Publications Last 3 Years:

1- Pelkonen O. Pasanen M., Tolonen A., Koskinen M., Hakkola J., Abass K., Laine J., Häkkinen M., Juvonen R., Auriola S., Storvik M., Huuskonen P., Rousu T., Rahikkala M. Reactive metabolites in early drug development: predictive in vitro tools, Current Medicinal Chemistry, *In press*
2- *Abass K. From in vitro hepatic metabolic studies towards human health risk assessment: A case study of diuron and carbosulfan. Pesticides Biochemistry and Physiology, 2014: 107(2) 258-256
3- *Abass K., Reponen P., Mattila S., Rautio A., Pelkonen O. Human variation and CYP enzyme contribution in benfuracarb metabolism in human in vitro hepatic models. Toxicology Letters, 2014: 224(2):300-9.

4- *Abass K., Reponen P., Mattila S., Rautio A., Pelkonen O. Comparative metabolism of benfuracarb in in vitro mammalian hepatic microsomes model and its implications for chemical risk assessment. Toxicology Letters, 2014: 224: 290-9.
5- *Abass K., Huusko A., Nieminen P., Myllynen P., Pelkonen O., Vahakangas K., Rautio A. Estimation of health risk by using toxicokinetic modelling: A case study of polychlorinated biphenyl PCB153. *Journal of Hazardous Materials.* 261 (2013) 1-10.
6- *Abass K., Pelkonen O. The inhibition of major human hepatic cytochrome P450 enzymes by 18 pesticides: Comparison of the N-in-one and single substrate approaches. *Toxicology In vitro.* 27 (2013) 1584-1588.
7- Pelkonen O., Abass K., Wiesner J. Thujone and thujone-containing herbal medicinal and botanical products: Toxicological assessment. *Regulatory Toxicology and Pharmacology.* 65 (2013) 100–107.
8- Pelkonen O., Turpeinen M., Hakkola J., Abass K., Pasanen M., Raunio H. How to preserve, induce or incorporate metabolism into the in vitro cellular system, *Toxicology in vitro.* 27 (2013) 1578–1583.

In: Biochemistry Research Summaries. Volume 1
Editors: James T. Walker and Catherine M. Jones
ISBN: 978-1-53610-895-8

Chapter 2

DEVENDRA K. AGRAWAL

Affiliation:

Professor of Biomedical Sciences, Internal Medicine, and Medical Microbiology and Immunology

Senior Associate Dean for Clinical & Translational Research Director, Center for Clinical & Translational Science

The Peekie Nash Carpenter Endowed Chair in Medicine

Contact Point:

CRISS II Room 510
Center for Clinical & Translational Science
Creighton University School of Medicine
2500 California Plaza, Omaha, NE 68178, USA Tel: (402) 280-2938
Fax: (402) 280-1421
E-mail: dkagr@creighton.edu

Date of Birth:

19 October 1953

Education:

2003 – 2005	Creighton University College of Business M.S. (Information Technology and Management)
2002 -2004	Creighton University College of Business M.B.A. (Business Administration)
1984 – 1985	University of British Columbia Faculty of Pharmaceutical Sciences Vancouver, Canada Postdoctoral Fellow (Pharmacology); Mentor: John H. McNeill, Ph.D.
1980 – 1984	McMaster University Department of Neurosciences Hamilton, Ontario, Canada Ph. D. (Medical Sciences); Mentor: Edwin E. Daniel, Ph.D.

1982 – 1984 McMaster University Faculty of Health Sciences Hamilton, Ontario, Canada
Postdoctoral Fellow (Physiology and Pharmacology) Mentor: Edwin E. Daniel, Ph.D.

1973 – 1978 King George's Medical College
Lucknow University
Postgraduate Department of Pathology & Bacteriology
Lucknow, U.P., India
Ph.D. (Biochemistry)
Mentors: Professor Abhaya Kumar, D. Phil. and
Professor R. M. L. Mehrotra, MBBS, MD, FRCPath

1971 – 1973 Lucknow University Department of Chemistry Lucknow, U.P., India
Master of Science (Chemistry) with specialization in
Organic Chemistry

1969 – 1971 Lucknow Christian College Lucknow University Lucknow, U.P., India
B. Sc. (Major in Zoology, Botany and Chemistry)

Professional Appointments:

07/2012 – present Senior Associate Dean for Clinical & Translational Research
Creighton University School of Medicine, Omaha, NE

04/2009 – 06/2012 Associate Dean for Translational Research
Creighton University School of Medicine, Omaha, NE

09/2009 – present Graduate Program Director Clinical & Translational Science
Creighton University, Omaha, NE

04/2009 – present Director
Center for Clinical and Translational Science
Creighton University School of Medicine, Omaha, NE

07/2004 – present Professor
Department of Biomedical Sciences (primary)
Department of Internal Medicine (secondary)
Department of Medical Microbiology & Immunology (secondary)
Creighton University School of Medicine, Omaha, NE

01/1998 – 06/2004 Professor
Department of Biomedical Sciences (secondary)
Creighton University School of Medicine, Omaha, NE

07/1997 – 06/2004 Professor
Department of Internal Medicine (primary)
Department of Medical Microbiology & Immunology (secondary)
Creighton University School of Medicine, Omaha, NE

07/1990 – 06/1997 Associate Professor (Tenured July 1990)
Department of Internal Medicine (primary)
Creighton University School of Medicine, Omaha, NE

07/1995 – 06/1997 Associate Professor
Department of Medical Microbiology & Immunology (secondary)
Creighton University School of Medicine, Omaha, NE

07/1990 – 06/1995 Associate Professor
Department of Pharmacology (secondary)
Creighton University School of Medicine, Omaha, NE

07/1990 – 06/1995 Associate Professor
Department of Biomedical Sciences (secondary)
Creighton University School of Medicine, Omaha, NE

07/1990 -06/2002 Director, Multidisciplinary Cardiovascular Research Group (formerly
Creighton Vascular Center), Creighton University Medical Center, Omaha, NE.

10/1985 – 06/1990 Assistant Professor
Department of Internal Medicine (primary)
Department of Pharmacology (secondary)
Creighton University School of Medicine, Omaha, NE

10/1985 – 06/2004 Director of Pre-clinical Research, Center for Allergy, Asthma and Immunology, Creighton University School of Medicine, Omaha, NE

09/1980 – 06/1982 Teaching Assistant (a resource person to M.D. students) Faculty of Health Sciences
McMaster University Medical Center
Hamilton, Ontario, Canada
Supervisor: Edwin E. Daniel, Ph.D.

10/1978 – 08/1980 Clinical Biochemist
King George's Medical College
Gandhi Memorial & Associated Hospitals
Lucknow, U.P., India

08/1973 – 08/1978	Research Assistant Postgraduate Department of Pathology and Bacteriology King George's Medical College, Lucknow, U.P. India Supervisors: Professor Abhaya Kumar and Professor RML Mehrotra

Honors:

11/2012	Invited Speaker at the 22nd National Annual Conference of the Indian Society of Hypertension in conjunction with the International Conference of Cardiovascular disease Prevention, Mumbai, India to speak on "Vitamin D as an Immunomodulator in Coronary Artery Disease"; 2-4 November 2012
09/2012	Invited Speaker at the Northwestern University School of Medicine, Chicago, IL to speak on the Molecular Mechanisms of Dendritic Cell Migration in Allergic Asthmatic Lungs; 6-7 September 2012
05/2012	Invited Speaker at the University of Pittsburgh Medical Center, Pittsburgh, PA to speak on Novel Mechanisms in the Pathogenesis of Chronic Asthma; 30 April-1 May 2012
01/2012	Invited Speaker at the 99th Indian Science Congress, Plenary session on Preventive Maternal and Child Health Care; Role of Vitamin D; Bhubaneswar, India, 3-9 January 2012
12/2011	Invited speaker at the 60th annual meeting of the American Society of Tropical Medicine & Hygiene, Philadelphia, PA, 4-8 December 2011
11/2011	Invited speaker at the mid-west regional meeting of the Network of Minority Research Investigators, NIH-NIDDK; November 3-4, 2011
06/2011	Invited speaker at the University of Nebraska Medical Center Immunology Interest Group; the title of my talk was "Novel Mechanisms in the Migration of Lung Dendritic Cells in Allergy and Asthma".
04/2011	Keynote Speaker at the Annual research Day of Creighton University; 5 April 2011; The title of the talk was "Multidisciplinary Knowledge Integration: Challenges and Rewards"
03/2011	Invited speaker and Co-chair at the luncheon seminar on "Th17 cells in Asthma and Autoimmunity"; 20 March 2011
02/2011	Invited Speaker at the annual meeting of the International Academy of Cardiovascular Sciences: Bridging the Gap: Basic Sciences and Clinical Practice, Vadodara, Gujarat, India, 1-3 February 2011
02/2011	Invited Speaker at the 4th World Congress of the International Academy of Cardiovascular Sciences: CIMS-3C-CON, Ahmedabad, Gujarat, India, 4-6 February 2011
01/2011	Invited Speaker at the 98th Indian Science Congress, Medical Sciences section; Chennai, India, 3-9 January 2011
12/2010	Invited Speaker at the Pittsburgh International Conference: LUNG CONFERENCE on the Understanding the Interface between Asthma, Host Defense, and Mucosal immunity, University of Pittsburgh, PA December 10-11, 2010
06/2010	Invited Speaker at the C. S. Medical University (formerly King Georg's Medical University), Lucknow, India; "Flt3-Ligand, Immunomodulation and Therapy in Asthma"

06/2010	Invited speaker at the Institute of Genomics and Integrative Biology, New Delhi, India; "Newer Treatments in Allergic Asthma"
03/2010	Invited Speaker at the annual meeting of the American Academy of Allergy, Asthma and Immunology, New Orleans, LO, in the Panel Discussion on "Who Dunnit: Which cells drive the asthma phenotype?"
07/2009	The Peekie Nash Carpenter Endowed Chair in Medicine
01/2009	Invited speaker, Wayne State University, Detroit, MI; "Vein Graft Disease; why is the internal mammary artery immune to restenosis?"
12/2008	Invited speaker in an International Symposium at the University of Vadodara, India on the Translational Research in Cardiovascular Medicine: From bench to bedside and vice-versa, Vadodara, India on December 11, 2008; "Vein-graft disease: Why is the internal mammary artery almost immune to restenosis?"
12/2008	Invited speaker in the 42nd Annual Conference of the Indian College of Allergy, Asthma and Applied Immunology, Ahmedabad, India on December 12, 2008; "Immunomodulation in the Therapy of Asthma".
12/2008	Invited speaker in the 42nd Annual Conference of the Indian College of Allergy, Asthma and Applied Immunology, Ahmedabad, India on December 13, 2008; "Are Eosinophils Pivotal in Allergic Diseases?"
12/2008	Invited speaker at the Delhi Institute of Pharmaceutical Sciences and Research, New Delhi, India in the Third International Symposium on "Recent Advances in Cardiovascular Sciences" on December 17, 2008; "Carotid Artery Disease: Why some people are symptomatic and others not?"
12/2008	Chair of a scientific session in the Third International Symposium on "Recent Advances in Cardiovascular Sciences", the Delhi Institute of Pharmaceutical Sciences and Research, New Delhi, India on December 17, 2008.
12/2008	Chair/Judge of the Poster session in International Conference on Translational Pharmacology, New Delhi, India on December 18, 2008.
12/2008	Invited speaker at the All India Institute of Medical Sciences, New Delhi, India, in the International Conference on Translational Pharmacology, on December 19, 2008; "Mechanisms underlying airway inflammation due to racemic β2-agonists".
11/2008	Invited speaker at the University of Virginia BB Carter center for Immunology Research, Charlottesville, VA on 3rd November 2008; "Flt3-Ligand, Immunomodulation and Therapy in Asthma".
09/2008	Invited Speaker in the Annual Research Forum, Sepracor Inc., New Orleans, LA; 18-20 September 2008; "Comparative effects of the enantiomers of formoterol on importins in human bronchial smooth muscle cells".
04/2008	Invited speaker at the University of Salzburg, Austria, in the 3rd International Symposium on Molecular Allergology – ISMA 2008 (Basic, Translational and Clinical Aspects), Salzburg, Austria, April 18-20, 2008.
03/2008	Invited speaker at the annual meeting of the American Academy of Allergy, Asthma and Immunology, Philadelphia, PA, March 13-18, 2008 to speak on "VEGF and Angiopoietins in Allergic Airway Inflammation and Airway Remodeling in Chronic Asthma".
03/2008	Invited speaker at the International Society for Heart Research – Indian section, held at the Postgraduate Institute of Medical Education and Research, Chandigarh, India, February 29- March 2, 2008.

02/2008	Invited speaker at the at the Delhi Institute of Pharmaceutical Sciences and Research, New Delhi, India to speak in the International symposium on Recent Advances in Cardiovascular Sciences, on February 28, 2008.
12/2007	Invited speaker at the annual meeting of the Indian Association of Clinical Biochemists, held in New Delhi, India, December 17-21, 2007.
08/2007	Invited speaker at the School of Medicine, St. Louis University in the Department of Medicine center for Immunobiology on 6th August 2007; "Novel Therapies in the Treatment of Allergic asthma".
03/2007	Invited speaker at the University of Nebraska Medical Center, Immunology Program on 14th March 2007; "Immunomodulation in Asthma".
11/2006	Invited speaker at the Cardiovascular Research Institute, Morehouse School of Medicine, Atlanta, GA; "Cellular and Molecular Mechanisms underlying Atherosclerotic Plaque Instability".
11/2006	Invited speaker at the International Symposium on Recent Research Advances in Asthma Pathogenesis, Grand Royal Antiguan Beach Resort, St. John's, Antigua, November 19-22, 2006.
10/2006	Invited speaker at the Biomedical Sciences Seminar Series, College of Veterinary Medicine, University of Minnesota, St. Paul, MN
03/2005	Invited speaker at the 3rd TAIHO Pharmaceutical Research Symposium, Pittsburgh, PA
10/2004	Chairman, A major symposium on Eosinophils at the 6th Asia-Pacific Congress of Allergology and Clinical Immunology, Tokyo, Japan.
10/2004	Invited speaker in a luncheon seminar, "Th1/Th2 Polarized Immunity" at the 6th Asia-Pacific Congress of Allergology and Clinical Immunology, Tokyo, Japan.
09/2004	Invited Speaker at the Sepracor, Inc. headquarters, Marlborough, MA
08/2004	Invited Speaker; University of Missouri, Columbia; Title of the talk was "Is the (S)-enantiomer of albuterol a silent bystander or culprit in worsening asthma?"
04/2004	Invited speaker at the 2nd TAIHO Pharmaceutical Research Symposium, Pittsburgh, PA.
01/2004	Invited Speaker at the TAIHO Pharmaceuticals, Tokushima Facility, Japan
01/2004	Invited Speaker, University of British Columbia, Vancouver, BC, Canada
09/2003	Invited speaker at the SEPRACOR Research Forum, New Orleans, LO.
03/2003	Invited speaker in a dinner symposium, "Can the β2-agonists be pro-inflammatory or anti-inflammatory agents?" at the annual meeting of the American Academy of Allergy, Asthma and Immunology, Denver, CO.
11/2002	Invited speaker in a dinner symposium, "Novel concepts in the treatment of allergic rhinitis" at the annual meeting of the American College of Allergy, Asthma and Immunology, San Antonio, TX
10/2002	Chairperson of an oral presentation session on "Asthma" at the 7th Asian Pacific Congress of Respirology, Taipei, Taiwan
10/2002	Invited Speaker in a Symposium on "Innovative Therapies in Asthma" at the 7th Asian Pacific Congress of Respirology, Taipei, Taiwan.
10/2001	Fellow, American Heart Association
10/2001	Fellow, Council of Arteriosclerosis, Thrombosis and Vascular Biology
03/2001	Fellow, American Academy of Allergy, Asthma and Immunology
02/2001	Invited speaker at the University of California, Davis, CA.

10/2000 Invited speaker at the 5th Tokyo Allergy Symposium, Tokyo, Japan

06/2000 Invited speaker at the Guidant Corporation, Santa Clara, CA.

11/1999 Invited speaker at the AtheroGenics, Alpharetta, GA.

03/1998 Invited speaker at the Genentech Corp, South San Francisco, CA, to speak on "Mechanisms underlying restenosis/intimal hyperplasia".

10/1996 Invited speaker at the IBC's 3rd International Conference on Restenosis, Cambridge, MA. The title of my talk was "Comparative effects of various stents on restenosis/intimal hyperplasia".

04/1992 Co-Chairperson of a session on "Vascular Smooth Muscle Pharmacology II" at the 1992 FASEB meeting, Anaheim, CA.

10/1991 Chairman of a session on monocyte/macrophage at the XIV International Congress of Allergology and Clinical Immunology, held in Kyoto, Japan, October 13-18, 1991.

02/1991 Chairman of a session on PAF Receptor Antagonists at the Tokyo PAF symposium on Allergic, Respiratory and Cardiovascular Diseases, held in Tokyo, Japan.

02/1991 Invited speaker at the Tokyo PAF symposium on allergic, respiratory and cardiovascular diseases, held in Tokyo, Japan.

05/1987 Invited speaker at the American Thoracic Society Annual meeting symposium on Bronchial Reactivity to speak on "Smooth Muscle: Modulation of Receptors and Response".

07/1985–10/1985 Postdoctoral fellow of the Canadian Heart Foundation to work on my research in the Faculty of Pharmaceutical Sciences, University of British Columbia, Vancouver, Canada.

07/1982–09/1984 Postdoctoral fellow of the Canadian Heart Foundation to work on my research project entitled "Alpha-adrenoceptors in vascular smooth muscle and their role in hypertension", at McMaster University, Hamilton, Ontario, Canada.

Publications Last 3 Years:

1. Makinde T, Murphy RF, Agrawal DK. The regulatory role of TGF-beta in airway remodeling in asthma. Immunol Cell Biol. In Press.
2. Moran EP, Agrawal DK. Increased expression of inhibitor of apoptosis proteins in atherosclerotic plaques of symptomatic patients with carotid stenosis. Exp Mol Pathol. In Press.

In: Biochemistry Research Summaries. Volume 1
Editors: James T. Walker and Catherine M. Jones
ISBN: 978-1-53610-895-8

Chapter 3

ALEKSANDRA ARSIC

Affiliation:

Centre of Research Excellence in Nutrition and Metabolism, Institute for Medical Research, University of Belgrade, Belgrade, Serbia

Contact Point:

dr Subotica 4a, Belgrade

Date of Birth:

22.11. 1971

Research and Professional Experience:

Aleksandra Arsic MSc, PhD, Research Associate is a biochemist. Since 2000, she works at University of Belgrade, Institute for medical research in Centre of Research Excellence in Nutrition and Metabolism. She is a scientist who is actively engaged in investigation of nutrition and metabolism and the role of lipids and fatty acids in healthy population and pathological conditions. One of the fields of research is the change in fatty acid composition of serum phospholipids in premature infants during neonatal life. Her interests also are the study of the effect of of fatty acid suplementation on plasma phospholipids fatty acid in different diseases. Her PhD thesis has the focus on the phospholipids fatty acid profile in serum and erythrocyte in female players and their oxidative status. Up to now, she has over 60 published papers, abstract and reports in national and international peer reviewed journals. She is member of a number professional associate in country: Physiological Society Serbia, Serbian Chemical Society, Biochemical Society of Serbia and the Serbian Society for mitochondrial and free-radical physiology.

Publications Last 3 Years:

1. Danijela Ristic-Medic, Perunicic-Pekovic Gordana, Zorica Rasic-Milutinovic, Marija Takic, Tamara Popovic, Aleksandra Arsic and Marija Glibetic "Effects of Dietary Milled Seed Mixture on Fatty Acid Status and Inflammatory Markers in Patients on Hemodialysis," The Scientific World Journal, vol. 2014, Article ID 563576, 9 pages, 2014. doi:10.1155/2014/563576.

2. Tepsic J, Vucic V, Arsic A, Mazic S, Delic M, Glibetic M. Unfavourable plasma and erythrocyte phospholipid fatty acid profile in elite amateur boxers. European Journal of Sport Science 2013; 13-414-423.
3. Arsić A, Vucić V, Tepsić J, Mazić S, Djelić M, Glibetić M. Altered plasma and erythrocyte phospholipid fatty acid profile in elite female water polo and football players. Applied Physiology Nutrition and Metabolism 2012; 37:40-47.
4. Popović T, Borozan S, Arsić A, Martačić JD, Vučic V, Trbović A, Mandić L, Glibetić M. Fish oil supplementation improved liver phospholipids fatty acid composition and parameters of oxidative stress in male wistar rats. Animal physiology and animal nutrition 2012; 96:1020–1029
5. Rasić-Milutinović Z, Popović T, Peruničić-Peković G, Arsić A, Borozan S, Glibetić M. Lower Serum Paraoxonase-1 Activity Is Related to Linoleic and Docosahexanoic Fatty Acids in Type 2 Diabetic Patients. Archives of Medical Research 2012; 43:75-82.
6. Popović T, Borozan S, Arsić A, Martačić JD, Vučic V, Trbović A, Mandić L, Glibetić M. Fish oil supplementation improved liver phospholipids fatty acid composition and parameters of oxidative stress in male wistar rats. Animal physiology and animal nutrition 2012; 96:1020–1029.
7. Arsic A, Vucic V, Prekajski N, Tepšic J, Ristic-Medic D, Velickovic V, Glibetic M. Different fatty acid composition of serum phospholipids of small and appropriate for gestational age preterm infants and of milk from their mothers. Hippokratia 2012; 16:230-235.

In: Biochemistry Research Summaries. Volume 1
Editors: James T. Walker and Catherine M. Jones ISBN: 978-1-53610-895-8

Chapter 4

ANTONIETTA BAIANO

Affiliation:

Dipartimento di Scienze Agrarie, degli Alimenti e dell'Ambiente, University of Foggia, Italy

Contact Point:

Via Napoli, 25 – 71122 Foggia, Italy

Date of Birth:

29/07/1972

Education:

Degree in Food Science and Technology, PhD in Food Biotechnology

Research and Professional Experience:

Researcher at the University of Foggia since 2002.

Publications Last 3 Years:

Baiano A., Antonacci D., Coletta A., Terracone C., La Notte E. - Effects of irrigation volume and nitrogen fertilization on quality of Redglobe and Michele Palieri table grape cultivars. American Journal of Enology and Viticulture, 62, 57-65, 2011.

BaianoA., Terracone C. - Dough rheology and bread quality of supplemented flours. Cyta, 9, 180-186, 2011.

Lamacchia C., Baiano A., Lamparelli S., Terracone C., Di Luccia A. - Study on the interactions between barley and semolina proteins during pasta-making and prediction of their effects on cooking behaviour and acceptability. Food Chemistry, 129, 319-328, 2011.

Baiano A., Terracone C. - Varietal differences among the phenolic profiles and antioxidant activities of seven table grape cultivars – Journal of Agricultural and Food Chemistry, 59, 9815-9826, 2011.

Baiano A., Terracone C. - Effects of bud load on quality of Beogradska and Thompson seedless table grapes and cultivars differentiation based on chemometrics of analitycal indices - Journal of the Science of Food and Agriculture, 92, 645-652, 2012.

Baiano A., Terracone C., Peri, G., Romaniello R. - Application of hyperspectral imaging for prediction of quality characteristics and sensory attributes of table grapes. Computers and Electronics in Agriculture, 142-151, 2012.

Fragasso M., Antonacci D., Pati S., Tufariello M., Baiano A., Forleo L.R., Caputo A., La Notte E. - Influence of training system on volatile and sensory profile of Primitivo grape and wine – presentato per la pubblicazione su American Journal of Enology and Viticulture, available online, doi: 10.5344/ajev.2012.11074.

Baiano A., Conte A., Del Nobile M.A. - Beer: Advances in Processing and Preservation - Journal Recent Patents on Engineering, 6, 2012.

Baiano A., Terracone C., Viggiani I., Del Nobile M.A. - Effects of cultivars and location on quality indices, phenolic content and antioxidant activity of extra-virgin olive oils –Journal of American Oil Chemists Society, in press.

Baiano A., Mastromatteo M., Del Nobile M.A. - Effects of cultivar and process variables on dynamic-mechanical and sensorial behaviour of added-value grape-based smoothies – Molecules, 17, 11421-11434, 2012.

Baiano A., Terracone C., Longobardi F., Ventrella A., Agostiano A., Del Nobile M.A. - Effects of different vinification technologies on physical and chemical characteristics of Sauvignon blanc wines - Food Chemistry, 135, 2694-2701, 2012.

In: Biochemistry Research Summaries. Volume 1
Editors: James T. Walker and Catherine M. Jones
ISBN: 978-1-53610-895-8

Chapter 5

WEI-HSIANG CHEN

Affiliation:
Institute of Environmental Engineering, National Sun Yat-sen University, Taiwan

Contact Point:
No.70, Lienhai Rd. Kaohsiung 80424, Taiwan

Date of Birth:
1977/07/26

Education:
Ph.D.

Research and Professional Experience:

Dr. Wei-Hsiang Chen is an Assistant Professor in the Institute of Environmental Engineering at National Sun Yat-Sen University (NSYSU) in Kaohsiung, Taiwan from August, 2011. Before joining NSYSU, Dr. Chen was a Post-Doctoral Scholar in John Muir Institute of the Environment at the University of California in Davis, CA (UCD) since January 2009. Dr. Chen received his Ph.D. degree of Environmental Engineering (Minor in Environmental Chemistry) from the Department of Civil and Environmental Engineering at UCD in 2009, his Master degree of Environmental Engineering, Science, and Management from the Department of Civil and Environmental Engineering at Carnegie Mellon University, Pittsburgh, PA in 2004, and his Bachelor degree from the Department of Water Resources and Environmental Engineering at Tamkang University in Taipei, Taiwan in 2000. Dr. Chen has publications in a number of peer reviewed journals and technical reports that include Environmental Science and Technology, Water Research, and Journal of Environmental Engineering by the American Society of Civil Engineers (ASCE). His research interests focus on the complex array of environmental physical and chemical processes and the associated applications to solve problems that pose risks to the environment and human health. Among many application domains in which environmental physical and chemical processes play fundamental roles, the lab is particularly interested in the areas of Water and Wastewater Treatment Technologies; Disinfection Byproducts and Emerging Contaminants; Fates and

Transports of Environmental Contaminants; Health Risk Assessment; and Sustainable Environmental Management and Strategies.

Professional Appointments:

Assistant Professor in NSYSU

Honors:

- Member of The Chinese Institute of Environmental Engineering
- Committee member of The Youth Committee in The Chinese Institute of Environmental Engineering
- Member of Taiwan Association for Aerosal Research
- Associate member in American Society of Civil Engineers
- Superior Student Paper Award, The 5th Cross-strait Environmental Protection Conference-Kaohsiung Forum (CEPC 2015), 2015/08
- Best Paper Award, 2015 Kaohsiung International Instruments Show, 2015/05
- Excellent Poster Paper Award, The 25th Annual Conference of The Chinese Institute of Environmental Engineering
- Excellent Paper Award, The 15th Mainland-Taiwan Environmental Protection Conference and 1st Resource, Environmental and Sustainable Development by Postgraduate's Academic Forum
- Excellent Poster Paper Award, The 23rd Annual Conference of The Chinese Institute of Environmental Engineering
- Student Travel Award, American Chemical Society, the 233rd National Meeting & Exposition, Chicago, MI, 2007

Publications Last 3 Years:

1. Wei-Hsiang Chen, Zheng-Bin Chen, and Chung-Shin Yuan* (November 7, 2015) "Investigating the differences between receptor and dispersion modeling for concentration prediction and health risk assessment of volatile organic compounds from petrochemical industrial complexes", *Journal of Environmental Management*, 166: 440-449, DOI: 10.1016/j.jenvman.2015.10.050 (SCIE; 2014 Impact Factor = 2.723; Ranking 58/221 in Environmental Sciences)
2. Wei-Hsiang Chen*, Ya-Ciao Yang, Ya-Hong Wang, Chi-Min Li, Kun-Yi Lin, and Jie-Chung Lou (October 22, 2015) "Effect of molecular characteristics on formation of nitrosamines during chlor(am)ination of phenylurea herbicides", *Environmental Science: Processes & Impacts*, 2015, 2092-2100, DOI: 10.1039/C5EM00308C (SCIE; 2014 Impact Factor = 2.171; Ranking 88/223 in Environmental Sciences)
3. Huazhen Shen, Iau-Ren Ie, Chung-Shin Yuan*, Chung-Hsuang Hung, Wei-Hsiang Chen, Jinjing Luo, Yi-Hsiu Jen (February 2015) "Enhanced photocatalytic oxidation of gaseous elemental mercury by TiO_2 in a high temperature environment", *Journal of Hazardous Materials*, 289, 235-243 (SCIE; 2014 Impact Factor = 4.529; Ranking 2/124 in Civil Engineering)
4. Jie-Chung Lou*, R.H. Lee, Wei-Hsiang Chen, Che-Jung, Chang; K.L. Hsu, Jia-Yun Han (November 2014) "Adsorption Kinetics and Thermodynamics of Perchlorate on

Carbon Nanotubes in Water", *Journal of Environmental Engineering, ASCE*, 140(11) 04014038 (SCIE; 2014 Impact Factor = 1.343; Ranking 40/124 in Civil Engineering)

5. Wei-Hsiang Chen, Yuan-Chung Lin*, Jun-Hong Lin*, Yang Po-Ming, and Jhang Syu-Ruei (October 2014) "Treating Odorous and Nitrogenous Compounds from Waste Composting by Acidic Chlorination followed by Alkaline Sulfurization", *Environmental Engineering & Science*, 31(11): 583-592. doi:10.1089/ees.2013.0272 (SCIE; 2014 Impact Factor = 0.991; Ranking 166/221 in Environmental Science)
6. Wei-Hsiang Chen*, Wen-Bin Yang, Chung-Shin Yuan*, Jun-Chen Yang, and Qing-Liang Zhao (May 2014) "Fates of Chlorinated Volatile Organic Compounds in Aerobic Biological Treatment Processes: The Effects of Aeration and Sludge Addition", *Chemosphere*, 103, 92-98 (SCIE; 2014 Impact Factor = 3.340; Ranking 39/221 in Environmental Science)
7. Yi-Hsiu Jen, Wei-Hsiang Chen, Chung-Shin Yuan*, Iau-Ren Ie, and Chung-Hsuang Hung (April 2014) "Seasonal Variation and Spatial Distribution of Atmospheric Mercury and Gas-particulate Partition in the Vicinity of a Semi-conductor Manufacturing Complex", *Environmental Science and Pollution Research*, 21(8), 5474-5483 (SCIE; 2014 Impact factor = 2.828; Ranking 54/221 in Environmental Sciences)
8. Jie-Chung Lou*, Che-Jung, Chang, Wei-Hsiang Chen, Wei-Bin Tseng, Jia-Yun Han (February 2014) "Removal of Trihalomethanes and Haloacetic Acids from Treated Drinking Water by Biological Activated Carbon Filter", *Water Air and Soil Pollution*, 225(2) 1851 (SCIE; 2014 Impact factor = 1.554; Ranking 35/83 in Water Resources)
9. Yi-Hsiu Jen, Wei-Hsiang Chen, Chung-Hsuang Hung, Chung-Shin Yuan*, and Iau-Ren Ie (February 2014) "Field Measurement of Total Gaseous Mercury and Its Correlation with Meteorological Parameters and Criteria Air Pollutants at a Coastal Site of the Penghu Islands", *Aerosol and Air Quality Research*, 14(1), 364-375 (SCIE; 2014 Impact Factor = 2.094; Ranking 93/221 in Environmental Sciences)
10. Chih-Chung Liu, Wei-Hsiang Chen*, Chung-Shin Yuan*, and Chitsan Lin (February 2014) "Multivariate Analysis of Effects of Diurnal Temperature and Seasonal Humidity Variations by Tropical Savanna Climate on the Emissions of Anthropogenic Volatile Organic Compounds", *Science of the Total Environment*, 470-471, 311-323 (SCIE; 2014 Impact Factor = 4.099; Ranking 18/221 in Environmental Sciences)
11. Iau-Ren Ie, HuaZhen She, Chung-Shin Yuan*, Chung-Hsuang Hung, Wei-Hsiang Chen, and Yi-Hsiu Jen (February 2014) "Single and Dual Adsorption of Vapor-phase Hg^0 and/or $HgCl_2$ by Innovative Composite Activated Carbons Impregnated with both S^0 and Na_2S", *Aerosol and Air Quality Research*, 14(1), 376-385 (SCIE; 2014 Impact Factor = 2.094; Ranking 93/221 in Environmental Sciences)
12. Junchen Yang, Kun Wang, Qingliang Zhao, Likun Huang, Chung-Shin Yuan, Wei-Hsiang Chen, and Wen-Bin Yang (February 2014) "Underestimated Public Health Risk Caused by Overestimated VOC Removal in Wastewater Treatment Processes", *Environmental Science: Processes and Impacts*, 16, 271-279 (SCIE; 2014 Impact Factor = 2.171; Ranking 88/221 in Environmental Sciences)
13. Yung-Chang Lin, Guo-Ping Chang-Chien, Pen-Chi Chiang, Wei-Hsiang Chen*, and Yuan-Chung Lin* (November 2013) "Multivariate Analysis of Heavy Metal

Contaminations in Seawater and Sediments from a Heavily Industrialized Harbor in Southern Taiwan", *Marine Pollution Bulletin*, 76(1-2), 266-275 (SCIE; 2014 Impact Factor = 2.991; Ranking 9/102 in Marine & Freshwater Biology)

14. Iau-Ren Ie, Chung-Hsuang Hung, Yi-Hsiu Jen, Chung-Shin Yuan*, and Wei-Hsiang Chen (August 2013) "Adsorption of Vapor-Phase Elemental Mercury (Hg^0) or Mercury Chloride ($HgCl_2$) with Innovative Composite Activated Carbons Impregnated with Na_2S and S^0 in Different Sequences", *Chemical Engineering Journal*, 229(1), 469-476 (SCIE; 2014 Impact Factor = 4.321; Ranking 9/134 in Chemical Engineering)
15. Yung-Chang Lin, Guo-Ping Chang-Chien, Pen-Chi Chiang, Wei-Hsiang Chen*, and Yuan-Chung Lin* (August 2013) "Potential Impacts of Discharges from Reverse Osmosis Seawater Desalination on the Marine Environment in Taiwan", *Desalination*, 322(1) 84-93 (SCIE; 2014 Impact Factor = 3.756; Ranking 2/83 in Water Resources)
16. Jhih-Jhe Yang, Chih-Chung Liu, Wei-Hsiang Chen*, Chung-Shin Yuan*, Chitsan Lin (May 2013) "Assessing the Altitude Effect on Distributions of Volatile Organic Compounds from Different Sources by Principal Component Analysis", *Environmental Science: Processes and Impacts*, 15 (5), 972-985 (SCIE; 2014 Impact Factor = 2.171; Ranking 88/221 in Environmental Sciences)
17. Tsung-Chang Li, Chung-Yi Wu, Wei-Hsiang Chen, Chung-Shin Yuan*, Shui-Ping Wu, Xin-Hong Wang, and Ke Du (April 2013) "Diurnal Variation and Chemical Characteristics of Atmospheric Aerosol Particles and Their Source Fingerprints at Xiamen Bay", *Aerosol and Air Quality Research*, 13(2), 596-607 (SCIE; 2014 Impact Factor = 2.094; Ranking 93/221 in Environmental Sciences)
18. Wei-Hsiang Chen*, Wen-Bin Yang, Chung-Shin Yuan*, Jun-Chen Yang, and Qing-Liang Zhao (February 2013) "Influences of Aeration and Biological Treatment on the Fates of Aromatic VOCs in Wastewater Treatment Processes", *Aerosol and Air Quality Research*, 13(1), 225-236 (SCIE; 2014 Impact Factor = 2.094; Ranking 93/221 in Environmental Sciences)
19. Tsung-Chang Li, Wei-Hsiang Chen, Chung-Shin Yuan*, Shui-Ping Wu, and Xin-Hong Wang (February 2013) "Physicochemical Characteristics and Source Apportionment of Atmospheric Particles in Kinmen-Xiamen Airshed", *Aerosal and Air Quality Research*, 13(1), 308-323 (SCIE; 2014 Impact Factor = 2.094; Ranking 93/221 in Environmental Sciences)
20. Wei-Hsiang Chen, Yuan-Chung Lin*, and Jun-Hong Lin (February 2013) "Removal of Odor-Causing Substances by A Pilot-Scale Oxidative-Reductive Scrubbing Tower System", *Aerosol and Air Quality Research*, 13(1), 385-395 (SCIE; 2014 Impact Factor = 2.094; Ranking 93/221 in Environmental Sciences)

In: Biochemistry Research Summaries. Volume 1
Editors: James T. Walker and Catherine M. Jones
ISBN: 978-1-53610-895-8

Chapter 6

Claudia Cocca

Affiliation:

Laboratorio de Efectos Biológicos de Contaminantes Ambientales, Departamento de Bioquímica Humana, Facultad de Medicina, Universidad de Buenos Aires, Buenos Aires, Argentina

Contact Point:

Junin 956. Ciudad Autonoma de Buenos Aires. Argentina

Date of Birth:

December 21st, 1965

Education:

Buenos Aires University

Research and Professional Experience:

Oncology and environmental risk factors; Environmental Toxicology

Professional Appointments:

Professor in Buenos Aires University, Buenos Aires, Argentina and established researcher of the CONICET.

Publications Last 3 Years:

Chlorpyrifos inhibits cell proliferation through ERK1/2 phosphorylation in breast cancer cell lines. Clara Ventura, Andrés Venturino, Noelia Miret, Andrea Randi, Elena Rivera, Mariel Núñez and Claudia Cocca. In Press to be published in Chemosphere (CHEMÂ 15316). Editorial Reference: CHEM_CHEM31755.

Action of hexachlorobenzene on tumor growth and metastasis in different experimental models. Pontillo CA, Rojas P, Chiappini F, Sequeira G, Cocca C, Crocci M, Colombo L, Lanari C, Kleiman de Pisarev D, Randi A. Toxicol Appl Pharmacol. 268(3):331-42 (2013).

Glibenclamide inhibits cell growth by inducing G0/G1 arrest in the human breast cancer cell line MDA-MB-231. Mariel Núñez, Vanina Medina, Graciela Cricco, Máximo Croci, Claudia Cocca, Elena Rivera, Rosa Bergoc, Gabriela Martín. BMC Pharmacol Toxicol 11;14:6 (2013).

Differential mechanisms of action are involved in chlorpyrifos effects in estrogen-dependent or -independent breast cancer cells exposed at low or high doses of the pesticide. Clara Ventura, Mariel Núñez, Noelia Miret, Diego Martinel Lamas, Andrea Randi, Andrés Venturino, Elena Rivera, Claudia Cocca. Toxicology Letters 213(2):184-93 (2012).

PANC-1 cells proliferative response to ionizing radiation is related to GSK-3β phosphorylation. Mohamad, Nora; Cricco, Graciela; Cocca, Claudia; Rivera, Elena; Bergoc, Rosa; Martín, Gabriela. Biochem Cell Biol. 90(6):779-90 (2012).

Alterations in c-Src/HER1 and Estrogen Receptor α signaling pathways in mammary gland and tumors of Hexachlorobenzene-treated rats. Peña D, Pontillo C, Garcia MA, Cocca C, Alvarez L, Chiappini F, Bourguignon N, Frahm I, Bergoc R, Kleiman de Pisarev D, Randi A. Toxicology 293(1-3):68-77 (2012). ISSN: 0300-483X.

In: Biochemistry Research Summaries. Volume 1
Editors: James T. Walker and Catherine M. Jones
ISBN: 978-1-53610-895-8

Chapter 7

GAMAL ESMAT

Affiliation:
Endemic Medicine and Hepatogastroenterology Department, Faculty of Medicine, Cairo University, Cairo, Egypt

Contact Point:
98 El Tahrir St., Bourg El- Dokki El-Edary, Dokki, Giza- Egypt

Date of Birth:
3rd Jan, 1956

Education:

1979	M.B.,B.Ch (Medical Degree) Cairo University Faculty of Medicine.
1984	MSc. (Master Degree in Tropical Medicine) Cairo University Faculty of Medicine.
1990	MD (Medical Doctorate in Tropical Medicine) Cairo University Faculty of Medicine.

Research and Professional Experience:
Completed Research Support

1. USAID and MOHPAbdel-Wahab (PI) 1989 - 1993
   Epidemiology of Schistosomiasis 1,2,3 sponsored by Schistosomiasis Research Project (SRP) Role: Co-PI
2. Schering Plough Sjogren (PI) 2001 -2003
   Efficacy of Interferon (PEG or alfa2-b) and Ribavirin in the Treatment of Chronic Hepatitis C in Egypt-Role: Local PI
3. Welcome Trust Strickland (PI) 7/1/00 - 6/30/06
   Evaluation of Protective, Disease-Associated & Therapeutic Immunity to Hepatitis C Virus Infection.
   Role: Director of Clinical Team

4. NIH-RO3 Gamal Esmat (PI) 2004-2007
Epidemiology of HCV Infection among Egyptian Children
Role: Principle Investigator
5. ANRS(12184) Gamal Esmat (PI) 2009-2010
Liver fibrosis evaluation among HCV genotype-4 infected patients in Egypt. Comparison of elastometry, histology and serum markers.
Role: Principle Investigator
6. Finland University Mohamed Abdel Hamid (PI) 2009-2011
Beneficial Health Effects of Probiotic Supplement to Hepatitis C Infected Patients
Role: Co-PI
7. Biogeneric Pharma Mohamed Karim (PI) 2010-2012
efficiency, dosing, safety and tolerability of Y-shaped interferon in chronic HCV patients
Role: Co-PI
8. Novartis Pharma Gamal Esmat (PI) 2010-2011
efficiency, dosing, safety of Sebivo (Telbuvidine) in treatment of chronic HBV
Role: Principle Investigator
9. BMS-AI444010 Gamal Esmat (PI) 2010-2012
Phase 2b trial of BMS-790052 in comibination with IFN and RBV in treatment of chronic HCV patients
Role: Principle Investigator
10. GSK NCT00529568 Gamal Esmat (PI) 2010-2011
Randomised, Placebo-controlled, Multi-centre Study to Assess the Efficacy and Safety of Eltrombopag in Thrombocytopenic Subjects With Hepatitis C Virus (HCV) Infection
Role: Principle Investigator

Ongoing Research Support

1. BMS-AI444042 Gamal Esmat (PI) 2012-2014
A Phase 3 Evaluation of BMS-790052 in Combination with Peg-Interferon Alfa-2a and Ribavirin in Treatment-Naive Subjects with Chronic Hepatitis C Genotype 4. Role: Principle Investigator

2. TMC435HPC3011—RESTORE STUDY Gamal Esmat (PI) 2012-2014
An open-label, single-arm Phase III study to evaluate the efficacy, safety and tolerability of TMC435 in combination with PegIFNÎ±-2a (Pegasys ®) and Ribavirin (Copegus ®) in treatment-naíve or treatment-experienced, chronic hepatitis C virus genotype-4 infected subjects. Role: Principle Investigator

Professional Appointments:

- President of IASL (International Association for the Study of the Liver) 2006-2008.
- Member of NCCVH & director of hepatitis treatment centers, NCCVH, MOHP(2006-now)

- Consultant of liver transplantation, DAF, Kasr Al Aini.
- WHO consultant for management of HBV.
- Director of schistosomal liver unit, Cairo University (2006-2010).
- Director of international research unit-NHTMRI, MOHP.
- Head of gastroenterology and hepatology unit, Al Kahera El Fatymia specialized medicine hospital.
- Director of abdominal ultra-sonography training program, Egyptian medical syndicate.
- Coordinatorof hepatogastroentrology post-graduate Courses, NTC, MOH.
- Director of liver unit , El Manial specialized hospital (2000-2004)
- Creator and Head of Kasr El Aini Viral Hepatitis Center (2010-till now).
- Vice president of Cairo university for graduate studies & research (2012- now)
- Member of Hepatitis B and C Summit
- Conferences Association, Luxembourg(2013-now)
- Board member in the Egyptian Society for Liver patinents.
- Egyptian medical society of Tropical Medicine, Parasitology and Infectious diseases,
- since1989.
- Egyptian medical society of Ultrasonography, since 1988.
- Egyptian medical society of Gastroenterology since 1989.
- The society of friends of liver patients in the Arab World since 1991.
- American Society of Tropical Medicine, since 1992.
- Mediterranean and African Association of Ultrasonography since 1999.
- Egyptian Medical Society of Hepatology since 1998.
- Permanent secretary of the African association of the study of liver disease (2003-2006).
- Fellow of the Maryland University at Baltimore
- Fellow of Edward Herriot hospital at Lyon France

Honors:

Academic Achievement Award: for 6 years from 1973 to 1979.
Graduation with honor Cairo University School of Medicine 1979.
Cairo Physician Excellency Award from Egyptian Medical Syndicate 1998.
President of the international liver congress, San Francisco (2008).
National Honorary Prize in Medicine (2010)

Publications Last 3 Years:

Mekky, Radwa Yehia, Nada Magdy El-Ekiaby, Mohammed Tarif Hamza, Noha Mousaad Elemam, Mohammed El-Sayed, Gamal Esmat, and Ahmed Ihab Abdelaziz. "Mir-194 is a hepatocyte gate keeper hindering HCV entry through targeting CD81 receptor." *The Journal of infection* (2014). Abstract

Breban, Romulus, Naglaa Arafa, Sandrine Leroy, Aya Mostafa, Iman Bakr, Laura Tondeur, Mohamed Abdel-Hamid, Wahid Doss, Gamal Esmat, Mostafa K. Mohamed et al. "Effect of preventive and curative interventions on hepatitis C virus transmission in Egypt

(ANRS 1211): a modelling study." *The Lancet. Global health* 2, no. 9 (2014): e541-9. Abstract

Saad, Yasmin, Olfat Shaker, Yasser Nassar, Lama Ahmad, Mohamed Said, and Gamal Esmat. "A polymorphism in the microsomal triglyceride transfer protein can predict the response to antiviral therapy in Egyptian patients with chronic hepatitis C virus genotype 4 infection." *Gut and liver* 8, no. 6 (2014): 655-61. Abstract

Elhelw, Dalia Sherif, Radwa Yehia Mekky, Nada El-Ekiaby, Rasha Ahmed, Mohammad Ahmed Mohey Eldin, Mohammad El-Sayed, Mahmoud Mohammad Abouelkhair, Ayman Salah, Abdel Rahman Zekri, Gamal Esmat et al. "Predictive prognostic role of miR-181a with discrepancy in the liver and serum of genotype 4 hepatitis C virus patients." *Biomedical reports* 2, no. 6 (2014): 843-848. Abstract

Alboraie, Mohamed, Marwa Khairy, Aisha Elsharkawy, Marwa Elsharkawy, Noha Asem, Amany R. Abo El-Seoud, Fathy G. Elghamry, and Gamal Esmat. "Egy-score as a noninvasive score for the assessment of hepatic fibrosis in chronic hepatitis C: A preliminary approach." *Saudi journal of gastroenterology: official journal of the Saudi Gastroenterology Association* 20, no. 3 (2014): 170-4. Abstract

Bruggmann, P., T. Berg, A. L. H. Øvrehus, C. Moreno, C. E. Brandão Mello, F. Roudot-Thoraval, R. T. Marinho, M. Sherman, S. D. Ryder, J. Sperl et al. "Historical epidemiology of hepatitis C virus (HCV) in selected countries." *Journal of viral hepatitis* 21 Suppl 1 (2014): 5-33. Abstract

Razavi, H., I. Waked, C. Sarrazin, R. P. Myers, R. Idilman, F. Calinas, W. Vogel, M. C. Mendes Correa, C. Hézode, P. Lázaro et al. "The present and future disease burden of hepatitis C virus (HCV) infection with today's treatment paradigm." *Journal of viral hepatitis* 21 Suppl 1 (2014): 34-59. Abstract

Wedemeyer, H., A. S. Duberg, M. Buti, W. M. Rosenberg, S. Frankova, G. Esmat, N. Örmeci, H. Van Vlierberghe, M. Gschwantler, U. Akarca et al. "Strategies to manage hepatitis C virus (HCV) disease burden." *Journal of viral hepatitis* 21 Suppl 1 (2014): 60-89. Abstract

Umar, Muhammed, Aamir G. Khan, Zaigham Abbas, Sanjeev Arora, Naqvi Asifabbas, Andre Elewaut, Gamal Esmat, Graham Foster, Michael Fried, Khean-L Goh et al. "World Gastroenterology Organisation global guidelines: diagnosis, management and prevention of hepatitis C April 2013." *Journal of clinical gastroenterology* 48, no. 3 (2014): 204-17. Abstract

Waked, Imam, Waheed Doss, Manal Hamdy El-Sayed, Chris Estes, Homie Razavi, Gamal Shiha, Ayman Yosry, and Gamal Esmat. "The current and future disease burden of chronic hepatitis C virus infection in Egypt." *Arab journal of gastroenterology: the official publication of the Pan-Arab Association of Gastroenterology* 15, no. 2 (2014): 45-52. Abstract

Bonnard, P., A. Elsharkawy, K. Zalata, E. Delarocque-Astagneau, L. Biard, L. Le Fouler, A. B. Hassan, M. Abdel-Hamid, M. El-Daly, M. E. Gamal et al. "Comparison of liver biopsy and noninvasive techniques for liver fibrosis assessment in patients infected with HCV-genotype 4 in Egypt." *Journal of viral hepatitis* (2014). Abstract

Hézode, Christophe, Gideon M. Hirschfield, Wayne Ghesquiere, William Sievert, Maribel Rodriguez-Torres, Stephen D. Shafran, Paul J. Thuluvath, Harvey A. Tatum, Imam Waked, Gamal Esmat et al. "Daclatasvir plus peginterferon alfa and ribavirin for treatment-naive chronic hepatitis C genotype 1 or 4 infection: a randomised study." *Gut* (2014). Abstract

Alboraie, Mohamed, Marwa Khairy, Marwa Elsharkawy, Noha Asem, Aisha Elsharkawy, and Gamal Esmat. "Value of Egy-Score in diagnosis of significant, advanced hepatic fibrosis and cirrhosis compared to aspartate aminotransferase-to-platelet ratio index, FIB-4 and Forns' index in chronic hepatitis C virus." *Hepatology research: The official journal of the Japan Society of Hepatology* (2014). Abstract

Esmat, Gamal, Maissa El Raziky, Aisha Elsharkawy, Dina Sabry, Mohamed Hassany, Amal Ahmed, Noha Assem, Mohamad El Kassas, and Wahid Doss. "Impact of Vitamin D Supplementation on Sustained Virological Response in Chronic Hepatitis C Genotype 4 Patients Treated by Pegylated Interferon/Ribavirin." *Journal of interferon & cytokine research: the official journal of the International Society for Interferon and Cytokine Research* (2014). Abstract

Fawzy, Injie O., Mohamed Negm, Rasha Ahmed, Gamal Esmat, Nabila Hamdi, and Ahmed I. Abdelaziz. "Tamoxifen downregulates MxA expression by suppressing TLR7 expression in PBMCs of males infected with HCV." *Journal of medical virology* 86, no. 7 (2014): 1113-9. Abstract

Esmat, Gamal, Mohamed El Kassas, Mohamed Hassany, Mohamed Gamil, and Maissa El Raziky. "Optimizing treatment for HCV genotype 4: PEG-IFN alfa 2a vs. PEG-IFN alfa 2b; the debate continues." *Liver international: official journal of the International Association for the Study of the Liver* 34 Suppl 1 (2014): 24-8. Abstract

Shehab, H., I. Elattar, T. Elbaz, M. Mohey, and G. Esmat. "CUFA algorithm: assessment of liver fibrosis using routine laboratory data." *Journal of viral hepatitis* 21, no. 12 (2014): 956-64. Abstract

Obach, Dorothée, Sylvie Deuffic-Burban, Gamal Esmat, Wagida A. Anwar, Sahar Dewedar, Valérie Canva, Anthony Cousien, Wahid Doss, Aya Mostafa, Stanislas Pol et al. "Effectiveness and cost-effectiveness of immediate versus delayed treatment of hepatitis C virus-infected patients in a country with limited resources: the case of Egypt." *Clinical infectious diseases: an official publication of the Infectious Diseases Society of America* 58, no. 8 (2014): 1064-71. Abstract

Cousien, Anthony, Dorothée Obach, Sylvie Deuffic-Burban, Aya Mostafa, Gamal Esmat, Valérie Canva, Mohamed El Kassas, Mohammad El-Sayed, Wagida A. Anwar, Arnaud Fontanet et al. "Is expert opinion reliable when estimating transition probabilities? The case of HCV-related cirrhosis in Egypt." *BMC Medical Research Methodology* 14 (2014): 39. Abstract

Ismail, Sohair, Hanan Abdel Hafez, Samar K. Darweesh, Kamal Hassan Kamal, and Gamal Esmat. "Virologic response and breakthrough in chronic hepatitis B Egyptian patients receiving lamivudine therapy." *Annals of gastroenterology: quarterly publication of the Hellenic Society of Gastroenterology* 27, no. 4 (2014): 380-386. Abstract 2013

Breban, R., W. Doss, G. Esmat, M. Elsayed, M. Hellard, P. Ayscue, M. Albert, A. Fontanet, and M. K. Mohamed. "Quantifying current hepatitis C virus incidence in Egypt - Response to letter by Miller and Abu-Raddad." *Journal of Viral Hepatitis* 20 (2013): 668. Abstract

Hatzakis, P. Van Damme, K. Alcorn, C. Gore, M. Benazzouz, S. Berkane, M. Buti, M. Carballo, Cortes H. Martins, S. Deuffic-Burban et al. "The State of Hepatitis B and C in the

Mediterranean and Balkan Countries: Report from a Summit Conference (vol 20, pg 1, 2013)." *Journal of Viral Hepatitis* 20 (2013): 744. Abstract

Abdel-Rahman, Mahasen, Yasmin Saad, Maissa El-Raziky, Naglaa Zayed, Wafaa El-Akel, Mohamed Said, Mohamed El-Beshlawy, and Gamal Esmat. "Hepatitis C genotype 4 with normal transaminases: Correlation with fibrosis and response to treatment, a cohort Egyptian study of 4277 patients." *Clinics and Research in Hepatology and Gastroenterology* 37 (2013): 479-484. Abstract

Khairy, Marwa, Maissa El-Raziky, Wafaa El-Akel, Mohamed S. Abdelbary, Hany Khatab, Badawy El-Kholy, Gamal Esmat, and Mahassen Mabrouk. "Serum autoantibodies positivity prevalence in patients with chronic HCV and impact on pegylated interferon and ribavirin treatment response." *Liver International* 33 (2013): 1504-1509. Abstract

Abdel-Rahman, Mahasen, Mohammad El-Sayed, Maissa El Raziky, Aisha Elsharkawy, Wafaa El-Akel, Hossam Ghoneim, Hany Khattab, and Gamal Esmat. "Coinfection with hepatitis C virus and schistosomiasis: Fibrosis and treatment response." *World Journal of Gastroenterology* 19 (2013): 2691-2696. Abstract

El Tayebi, H. M., K. Omar, S. Hegy, M. El Maghrabi, M. El Brolosy, K. A. Hosny, G. Esmat, and A. I. Abdelaziz. "Repression of miR-17-5p with elevated expression of E2F-1 and c-MYC in non-metastatic hepatocellular carcinoma and enhancement of cell growth upon reversing this expression pattern." *Biochemical and Biophysical Research Communications* 434 (2013): 421-427. Abstract

Raziky, Maissa El, Waleed Fouad Fathalah, Wafaa Ahmed El-akel, Ahmed Salama, Gamal Esmat, Mahassen Mabrouk, Rabab Mamoun Salama, and Hany Mahmoud Khatab. "The Effect of Peginterferon Alpha-2a vs. Peginterferon Alpha-2b in Treatment of Naive Chronic HCV Genotype-4 Patients: A Single Centre Egyptian Study." *Hepatitis Monthly* 13 (2013). Abstract

El-Kady, Nabeel M., Gamal Esmat, Ekram H. Mahmoud, Samar K. Darweesh, Sherif H. Mahmoud, and Waleed A. Elagawy. "Hypertonic saline-enhanced radiofrequency versus chemoembolization sequential radiofrequency in the treatment of large hepatocellular carcinoma." *European Journal of Gastroenterology and Hepatology* 25 (2013): 628-633. Abstract

Esmat, Gamal. "Hepatitis C in the Eastern Mediterranean Region." *Eastern Mediterranean Health Journal* 19 (2013): 587-588. Abstract

Elalfy, Mohsen S., Gamal Esmat, Randa M. Matter, Hesham Abdel E. Aziz, and Walid A. Massoud. "Liver fibrosis in young Egyptian beta-thalassemia major patients: relation to hepatitis C virus and compliance with chelation." *Annals of Hepatology* 12 (2013): 54-61. Abstract

Esmat, Gamal, Mohamed El Kassas, Mohamed Hassany, Mohamed Esmat Gamil, and Maisa El Raziky. "How to optimize HCV therapy in genotype 4 patients." *Liver International* 33 (2013): 41-45. Abstract

Ragab, Dina, Melissa Laird, Darragh Duffy, Armanda Casrouge, Rasha Mamdouh, Amal Abass, Dina El Shenawy, Abdelhadi M. Shebl, Wagdi F. Elkashef, Khaled R. Zalata et al. "CXCL10 antagonism and plasma sDPPIV correlate with increasing liver disease in chronic HCV genotype 4 infected patients." *Cytokine* 63 (2013): 105-112. Abstract

Assal, R., Mohamed H. El Tayebi, Adel K. Hosny, G. Esmat, and Ihab A. Abdelaziz. "Failure of induction of the oncomiRic cluster miR-182/miR-96 in HCV-induced non-metastatic hepatocellular carcinoma tissues renders them as markers for metastasis." *Virchows Archiv* 463 (2013): 125. Abstract

El-Kamary, Samer S., Mona M. Mohamed, Maissa El-Raziky, Michelle D. Shardell, Olfat G. Shaker, Wafaa A. ElAkel, and Gamal Esmat. "Liver fibrosis staging through a stepwise analysis of non-invasive markers (FibroSteps) in patients with chronic hepatitis C infection." *Liver International* 33 (2013): 982-990. Abstract

Habashy, D., Mohamed H. El Tayebi, Adel K. Hosny, G. Esmat, and Ihab A. Abdelaziz. "MicroRNA-17-5p promotes tumorgenesis through the repression of IGFBP-3 and enhancement of insulin like growth factor signaling." *Virchows Archiv* 463 (2013): 322. Abstract

Tayel, Sara S., Amal A. Helmy, Rasha Ahmed, Gamal Esmat, Nabila Hamdi, and Ahmed Ihab Abdelaziz. "Progesterone suppresses interferon signaling by repressing TLR-7 and MxA expression in peripheral blood mononuclear cells of patients infected with hepatitis C virus." *Archives of Virology* 158 (2013): 1755-1764. Abstract

Hatzakis, A., P. Van Damme, K. Alcorn, C. Gore, M. Benazzouz, S. Berkane, M. Buti, M. Carballo, H. Cortes Martins, S. Deuffic-Burban et al. "The State of Hepatitis B and C in the Mediterranean and Balkan Countries: Report from a Summit Conference." *Journal of Viral Hepatitis* 20 (2013): 1-20. Abstract

Mohsen, A., E. M. Ibrahim, S. Al Aidi, S. El Mango, A. Bernier, M. El Daly, M. El Gafaary, Abdel M. Hamid, G. Esmat, M. Gadallah et al. "Community Transmission of Hepatitis C Virus among Egyptians: Analysis of a 10 Years Case-Control Study." *Journal of Hepatology* 58 (2013): S398. Abstract

El-Sayed, M., M. El-Raziky, A. Elsharkawy, H. Khattab, A. H. K. Abdelmaksoud, M. K. Ashour, and G. Esmat. "Comparison between Diffusion-Weighted Mri, Fibroscan and Histopathology for Assessment of Liver Fibrosis in Chronic Hcv Patients." *Journal of Hepatology* 58 (2013): S185. Abstract

Bonnard, P., M. El-Kassas, M. Gamal, A. B. Hassan, L. Le Fouler, A. Elsharkawy, E. Delarocque-Astagneau, M. El-Sayed, A. Abd El Hay, M. Abdel-Hamid et al. "Comparison of Liver Biopsy, Elastometry and Serum Markers for Liver Fibrosis Assessment in Genotype 4 Hcv-Infected Patients in Egypt. Results of the ANRS12184 study." *Journal of Hepatology* 58 (2013): S287. Abstract

Zekri, A. - R., H. Shousha, A. Nassar, N. Zayed, G. Esmat, A. Abdel-Aziz, and A. Kaseb. "Gene Methylation Markers for Egyptian Patients with Hepatocellular Carcinoma." *Journal of Hepatology* 58 (2013): S256. Abstract

Waly, A., H. M. El Tayebi, K. A. Hosny, G. Esmat, and A. I. Abdelaziz. "Hypermethylation of the microRNA LET-7A-3 Gene Represses the Primary and Mature LET-7A-3 with an Inverse Correlation to IGF-II mRNA in Hcv-Induced Hepatocellular Carcinoma." *Journal of Hepatology* 58 (2013): S448-S449. Abstract

Obach, D., S. Deuffic-Burban, G. Esmat, W. A. Anwar, S. Dewedar, V. Canva, A. Cousien, W. Doss, A. Mostafa, S. Pol et al. "Immediate vs. Delayed Treatment in GenotypE 4-HCV Infected Patients in a Limited Resources Country: A Cost-Effectiveness Analysis in Egypt (ANRS 12215)." *Journal of Hepatology* 58 (2013): S364. Abstract

Breban, R., N. Arafa, S. Leroy, A. Mostafa, I. Bakr, M. Adbel-Hamid, G. Esmat, M. K. Mohamed, and A. Fontanet. "Impact of Mass Antiviral Treatment on HCV Transmission in Rural Egypt." *Journal of Hepatology* 58 (2013): S23. Abstract

El Raziky, M., A. B. Awad, M. Youssef, A. Awad, A. Elsharkawi, G. Esmat, M. Mustafa, and M. Mabrouk. "A Novel Prediction Model for Liver Fibrosis in Patients with Chronic Hepatitis C Virus Using Fibroscan and Routine Laboratory Data." *Journal of Hepatology* 58 (2013): S184. Abstract

Breban, R., W. Doss, G. Esmat, M. Elsayed, M. Hellard, P. Ayscue, M. Albert, A. Fontanet, and M. K. Mohamed. "Towards realistic estimates of HCV incidence in Egypt." *Journal of Viral Hepatitis* 20 (2013): 294-296. Abstract

Mabrouk, Mahasen, Maissa El-Raziky, Naglaa Zayed, Rabab Salama, Wafaa El-Akel, Tahany Awad, Mohammad El Beshlawy, and Gamal Esmat. "Clinical, biochemical and pathological profiles of 5464 Egyptian chronic hepatitis C-infected patients." *Hepato-gastroenterology* 60, no. 127 (2013): 1731-5. Abstract

Shaker, Olfat, Heba Bassiony, Maissa El Raziky, Samer S. El-Kamary, Gamal Esmat, Akmal M. El-Ghor, and Mona M. Mohamed. "Human Leukocyte Antigen Class II Alleles (DQB1 and DRB1) as Predictors for Response to Interferon Therapy in HCV Genotype 4." *Mediators of Inflammation* (2013).

In: Biochemistry Research Summaries. Volume 1
Editors: James T. Walker and Catherine M. Jones
ISBN: 978-1-53610-895-8

Chapter 8

Pablo Andres Evelson

Affiliation:

Quimica General e Inorganica, Instituto of BioquÃmica y Medicina Molecular (IBIMOL-UBA-CONICET), Facultad de Farmacia y BioquÃmica. Universidad de Buenos Aires, Buenos Aires, Argentina.

Contact Point:

Quimica General e Inorganica,
Facultad de Farmacia y BioquÃmica.
Universidad de Buenos Aires,
Junin 956 -2do Piso
1113 - Ciudad AutÃ³noma de Buenos Aires
Argentina.

Date of Birth:

March 13, 1967

Education:

Institution: University of Buenos Aires. Argentina.
Biochemist (1991). School of Pharmacy and Biochemistry. University of Buenos Aires.
Ph.D. in Biophysics (1998). School of Pharmacy and Biochemistry.
University of Buenos Aires.

Research and Professional Experience:

-1988-1990: Laboratory Instructor in Physical Chemistry. Department of Biological Chemistry. School of Pharmacy and Biochemistry. University of Buenos Aires.

-1991-1992: Instructor in Physical Chemistry. Department of Analytical and Physical Chemistry. School of Pharmacy and Biochemistry. University of Buenos Aires.

-1992-1993: Research Fellow. Humana Hospital Michael Reese. University of Illinois at Chicago. Chicago, Illinois, USA.

-1993-1998: Instructor in Physical Chemistry. School of Pharmacy and Biochemistry. University of Buenos Aires.

-1998-1999: Posdoctoral Fellow. Physiology Program. Department of Environmental Health. Harvard School of Public Health. Harvard University. Boston, Massachusetts, USA.

-2000-2005: Instructor in General and Inorganic Chemistry. School of Pharmacy and Biochemistry. University of Buenos Aires.

-2005-2012: Assistant Professor in General and Inorganic Chemistry. School of Pharmacy and Biochemistry. University of Buenos Aires.

-2013: Associate Professor in General and Inorganic Chemistry. School of Pharmacy and Biochemistry. University of Buenos Aires.

Professional Appointments:

-Associate Professor in General and Inorganic Chemistry. School of Pharmacy and Biochemistry. University of Buenos Aires.

-Career Investigator. National Council of Science and Technology (CONICET)

Honors:

-1992: Research Fellowship. Humana Hospital Michael Reese. University of Illinois at Chicago. Chicago, Illinois, USA.

-1993: Young Investigator Award. School of Pharmacy and Biochemistry. University of Buenos Aires. Argentina.

-1993-1998: Doctoral Fellowship. University of Buenos Aires. Argentina.

-2003-2004: Postdoctoral Fellowship. National Council of Science and Technology (CONICET). Argentina.

-2005: Adolfo Nocetti - Alberto Tiscorn

Publications Last 3 Years:

Vanasco V, Magnani N, Cimolai MC, Valdez L, Evelson P, Boveris A, Alvarez S. 2012. Endotoxemia impairs heart mitochondrial function by decreasing electron transfer, ATP synthesis and ATP content without affecting membrane potential. *Journal of Bioenergetics and Biomembranes*. 44: 243-252.

Busso Casati C, Sánchez V, Baeza Rosa, Magnani N, Evelson P, Zamora MC. 2012. Relationships between color parameters, phenolic content and sensory changes of processed blueberry, elderberry and blackcurrant commercial juices. *International Journal of Food Science & Technology*. 47:1728-1736.

Marchini T, Magnani N, D'Annunzio V, Tasat D, Gelpi R, Alvarez S, Evelson P. 2013. Impaired cardiac mitochondrial function and contractile reserve following an acute exposure to environmental particulate matter. *Biochimica et Biophysica Acta*. 1830: 2545-2552.

Leal MC, Magnani N, Villordo S, Marino Buslje C, Evelson P, Castaño E, Morelli L. 2013. Transcriptional regulation of insulin degrading enzyme modulates mitochondrial AÎ² catabolism and functionality. *Journal of Biological Chemistry*. 288: 12920-12931.

Magnani N, Marchini T, Vanasco V, Tasat D, Alvarez S, Evelson P. 2013. Reactive oxygen species produced by NADPH oxidase and mitochondrial dysfunction in lung after an acute exposure to Residual Oil Fly Ashes. *Toxicology and Applied Pharmacology*. 270: 31–38.

Calero C, Beltrán González A, Gasulla J, Alvarez S, Evelson P, Calvo D. 2013. Quercetin antagonism of GABAAρ1 receptors is prevented by ascorbic acid through a redox-independent mechanism. *European Journal of Pharmacology*. 714: 274-280.

Marchini T, Magnani ND, Paz ML, Vanasco V, Tasat D, González Maglio DH, Alvarez S, Evelson P. 2014. Time course of systemic oxidative stress and inflammatory response induced by an acute exposure to Residual Oil Fly Ash. *Toxicology and Applied Pharmacology*. 274: 274-282.

Bruno ME, Tasat D, Ramos E, Paparella ML, Evelson P, Rebagliatti R, Cabrini R, Guglielmotti M, Olmedo D. 2014. Impact through time of different sized titanium dioxide particles on biochemical and histopathological parameters. *Journal of Biomedical Materials Research: Part A*. 102: 1439-1448.

Vanasco V, Saez T, Magnani ND, Pereyra L, Marchini T, Corach A, Vaccaro MI, Corach D, Evelson P, Alvarez S. (2014) Cardiac mitochondrial biogenesis in endotoxemia is not accompanied by mitochondrial function recovery. *Free Radical Biology & Medicine*. 277:1-9.

Cimolai MC, Vanasco V, Marchini T, Magnani N, Evelson P, Alvarez S. (2014) α-lipoic acid protects kidney from oxidative stress associated to inflammatory conditions. *Food & Function*. 5: 3143-3150.

In: Biochemistry Research Summaries. Volume 1
Editors: James T. Walker and Catherine M. Jones
ISBN: 978-1-53610-895-8

Chapter 9

JIAN-LI GAO

Affiliation:
Zhejiang Chinese Medical University

Contact Point:
No. 548. Binwen Road, Binjiang District, Hangzhou, Zhejiang 310053, China

Date of Birth:
Jan 12th, 1982

Education:
Ph.D. in Biomedical Sciences

Research and Professional Experience:
Sep. 2013 ~ present: Research Associate of Zhejiang Chinese Medical University, Hangzhou, China.

Aug. 2011 ~ Sep. 2013: Research assistant of Zhejiang Chinese Medical University, Hangzhou, China.

Nov. 2009 ~ Jun. 2011: Postdoctoral Fellow in Molecular Pharmacology, Molecular Oncology Laboratory, The University of Chicago Medical center, Chicago, IL

Professional Appointments:
Research Associate

Publications Last 3 Years:

1) He Kai, Gao JL*. Protopine inhibits heterotypic cell adhesion in MDA-MB-231 cells through down-regulation of multi-adhesive factors. African Journal of Traditional, Complementary and Alternative Medicines. 2014. 11(2): 415-424. (Corresponding Author) IF2013= 0.560

2) Zhao Y, Gao JL*, Ji JW, Gao M, Yin QS, Qiu QL, Wang C, Chen SZ, Xu J, Liang RS, Cai YZ, Wang XF. Cytotoxicity enhancement in MDA-MB-231 cells by the combination treatment of tetrahydropalmatine and berberine derived from Corydalis yanhusuo W. T. Wang. Journal of Intercultural Ethnopharmacology. 2014; 3(2): 68-72 (Corresponding Author).

3) Gao JL, Lv GY, He BC, Zhang BQ, Zhang H, Wang N, Wang CZ, Du W, Yuan CS, He TC. Ginseng saponin metabolite 20(S)-protopanaxadiol inhibits tumor growth by targeting multiple cancer signaling pathways. Oncol Rep. 2013, 30(1): 292-8. IF2013= 2.191

4) Gao JL, Ji X, He TC, Zhang Q, He K, Zhao Y, Chen SH, Lv GY. Tetrandrine suppresses cancer angiogenesis and metastasis in 4T1 tumor-bearing mice. Evid Based Complement Alternat Med. 2013, Article ID 265061, 12 pages. IF2013= 2.175

5) Lv GY, Zhang YP, Gao JL, Yu JJ, Lei J, Zhang ZR, Li B, Zhan RJ, Chen SH. Combined Antihypertensive Effect of Luteolin and Buddleoside Enriched Extracts in Spontaneously Hypertensive Rats. J ETHNOPHARMACOL. 2013 Nov 25;150(2):507-13. doi: 10.1016/j.jep.2013.08.058. Epub 2013 Sep 28. (Corresponding Author) IF2013= 2.939

In: Biochemistry Research Summaries. Volume 1
Editors: James T. Walker and Catherine M. Jones
ISBN: 978-1-53610-895-8

Chapter 10

BOJAN Ž. JANKOVIĆ

Affiliation:

Faculty of Physical Chemistry, Department for Dynamics and Matter Structure, University of Belgrade, Studentski trg 12-16, P. O. Box 137, 11001 Belgrade, Serbia

Contact Point:

Belgrade, Studentski trg 12-16

Date of Birth:

10. October 1974

Education:

Ph.D.

Research and Professional Experience:

Solid State Kinetics, Statistical methods and distributions, Polymer degradation kinetics and thermal stability, Biopolymers, Applied Energy, Energy and Fuels, Pyrolysis and modeling, Metallurgy – Physical Metallurgy, Pharmaceuticals – Storage and Thermal stability of active components of drugs, Catalysis, Alloys, Inorganic composites and Polymer composites, Intermetallics, Thermodynamics, Survival Analysis.

Professional Appointments:

Professor, Educations of students, Expert Advisor at RWTH Aachen University (Germany), in IME Process Metallurgy and Metal Recycling Center (Aachen, Germany).

The active participant in the International Project FR 1713/13-1 titled "Zn recovery from steel making dusts - Kinetics and mechanism of thermal zinc-ferrite phase decomposition", financed by the German government (Deutsche Forschungsgemeinschaft DFG, Bonn, Germany).

Member of the Serbian Physical Chemistry Society, and the Serbian Chemical Society.

Honors:

The standing Reviewer of a more than 30 prominent International Journals worldwide. In the home country: "Pavle Savic" Award for the most successful completed University studies.

Publications Last 3 Years:

Bojan Janković, Milena Marinović-Cincović, Vojislav Jovanović, Suzana Samarţija-Jovanović, Gordana Marković, The comparative kinetic analysis of non-isothermal degradation process of acrylonitrile–butadiene and ethylene–propylene–diene rubber compounds. Part I, *Thermochimica Acta*, **543**, 2012, 295-303. (IF) 1.908, 33/73;

Bojan Janković, Srećko Stopić, Aybars Guven, Bernd Friedrich, The application of the formalism of dispersive kinetics for investigation of the isothermal decomposition of zinc leach residue in an inert atmosphere *Thermochimica Acta*, **546**, 2012, 102-112. (IF) 1.908, 33/73;

Bojan Janković, Milena Marinović-Cincović, Vojislav Jovanović, Suzana Samardzija-Jovanović, Gordana Marković, The comparative kinetic analysis of non-isothermal degradation process of acrylonitrile–butadiene/ethylene–propylene–diene rubber blends reinforced with carbon black/silica fillers. Part II, *Thermochimica Acta*, **543**, 2012, 304-312. (IF) 1.908, 33/73;

Bojan Janković, Kinetic Modeling of Native *Cassava* Starch Thermooxidative Degradation Using Weibull and Weibull-Derived Models, *Biopolymers* 'Accepted Article', doi: 10.1002/bip.22271.(IF) 2.870, 32/74;

Bojan Janković, Srećko Stopić, Aybars Güven, Bernd Friedrich, Kinetic Analysis of Isothermal Decomposition Process of Zinc Leach Residue in an Inert Atmosphere: The Estimation of the Apparent Activation Energy Distribution *Mineral Processing & Extractive Metallurgy Review* Accepted author version posted online: 11 Apr 2013 **DOI:** 10.1080/08827508.2013.766793. (IF) 0.667, 9/23;

B. Janković, Identification of the effective distribution function for determination of the distributed activation energy models using Bayesian statistics: Application of isothermal thermogravimetric data, *International Journal of Chemical Kinetics*, **42**, 2010, 641-658. (IF) 1.154, 93/127;

B. Janković, Thermal stability investigation and the kinetic study of Folnak® degradation process under nonisothermal conditions, *AAPS PharmSciTech* (An Official Journal of the American Association of Pharmaceutical Scientists), **11**, 2010, 103-112. (IF) 1.211, 190/252;

B. Janković, S. Mentus, Thermal stability and nonisothermal kinetics of Folnak® degradation process, *Drug Development and Industrial Pharmacy*, **36**, 2010, 980-992. (IF) 1.396, 37/54;

Marija M. Janković, Bojan Ž. Janković, Dragana J. Todorović, Ljubiša M. Ignjatović, Tritium concentration analysis in atmospheric precipitation in Serbia, *Journal of Environmental Science and Health Part A*, **47**, 2012, 669-674. (IF) 1.190, 131/205;

Bojan Ž. Janković, Marija M. Janković, Kinetic and thermodynamic analysis of creosote degradation process under isothermal experimental conditions, *Journal of Environmental Science and Health Part A*, **48**, 2013, 1437-1449. (IF) 1.190, 131/205;

B. Janković, The kinetic analysis of sodium bicarbonate ($NaHCO_3$) decomposition process using nonisothermal approach, *The 42nd International October Conference on Mining and Metallurgy IOC 2010 Proceedings*, October 10-13, Kladovo, Serbia, University of Belgrade, Technical Faculty in Bor and Mining and Metallurgy Institute, Bor, 2010, 521-524.

Bojan Ž. Janković, Kinetics of isothermal pyrolysis of pine wood. Determination of reaction mechanism. The fifth international conference "Ecology, health, work, sport" Banja Luka, Book of proceedings, Ed. S. Dikić, 6 - 9 september 2012., Banja Luka, Republic of Srpska, p. 556-561. ISBN 987-99955-789-3-6.

M. M. Janković, D. J. Todorović, B. Ž. Janković, J. D. Nikolić, G. K. Pantelić, M. M. Rajačić, Beryllium-7 concentration analysis in ground level air in Serbia. Physical Chemistry 2012, Proceedings of the 11th International Conference on Fundamental and Applied Aspects of Physical Chemistry, Ed. S. Anić, Ž. Čupić, September 24-28, Society of Physical Chemists of Serbia, p. 639-641, Belgrade, 2012. ISBN 978-86-82475-28-6.

Bojan Janković, Srećko Stopić, Bernd Friedrich, Phase transformation kinetics during isothermal decomposition of iron-zinc concentrate in nitrogen atmosphere. 6th Symposium Chemistry and Environmental Protection with international participation, Book of Abstracts, Vršac, May 21-24. 2013. p. 138-139. Serbian Chemical Society, Eds. Ivan Gržetić, Bojan Radak, Vladimir Beškovski, ISBN: 978-86-7132-052-8.

In: Biochemistry Research Summaries. Volume 1
Editors: James T. Walker and Catherine M. Jones
ISBN: 978-1-53610-895-8

Chapter 11

RAMAZ KATSARAVA

Affiliation:
Institute of Chemistry and Molecular Engineering, Agricultural University of Georgia

Contact Point:
28, Janashia str., ap.26, Tbilisi 0179, Georgia

Date of Birth:
15 September 1943

Education:

1966 – graduated with first-class honors from Georgian Polytechnic Institute (Tbilisi),

1969 – post-graduated from D.I.Mendeleev Institute of Chemical Technology (Moscow),

1971- Candidate of Chemical Sci. (Ph.D.), D.I. Mendeleev Institute of Chemical Technology (Moscow),

1988 – Doctor of Chemical Sci. A.N. Nesmeyanov Institute of Element-Organic Compounds (INEOC) of Academy of Sciences of USSR (Moscow).

Research and Professional Experience:

New Polycondensation (PC) methods of polymer synthesis (Active Polycondensation). Synthesis and study of novel biodegradable and biocompatible polymers composed of naturally occurring α-amino acids. Biomedical applications of the new biodegradable polymers as resorbable surgical devices, sustained/controlled drug eluting systems, drug-delivering microspheres and nanoparticles, etc.

Professional Appointments:

1970-1973 – researcher, A.N.Nesmeyanov Institute of Element-organic Compounds (INEOC) of Russian Academy of Sciences (Moscow),

1974-1986 – senior researcher, Inst. of Physiology of the Georgian Acad. of Sci. (Tbilisi),

1986-1998 – Head of the Laboratory, Inst. of Molecular Biology and Biophysics of the Georgian Acad. of Sci. (Tbilisi),

1993 to date – Professor of the Georgian Technical University (GTU, Tbilisi),

1997 to date – Head of the Research Centre for Medical Biotechnology & Bioengineering - CRDF/RESCenter at GTU,

2009-2013 – Director of the Institute of Medical Polymeric Materials, I. Javakhishvili Tbilisi State University,

2013 to date - Director of the Institute of Chemistry and Molecular Engineering, Agricultural University of Georgia.

Honors:

Honour's order-bearer of Georgia (1998),

World Intellectual Property Organization's (WIPO) Gold Medal (2007)

Publications Last 3 Years:

N. Zavradashvili, G. Jokhadze, M. Gverdtsiteli, G. Otinashvili, N. Kupatadze, Z. Gomurashvili, D. Tugushi, R. Katsarava, Amino Acid Based Epoxy-Poly(Ester Amide)s - a New Class of Functional Biodegradable Polymers: Synthesis and Chemical Transformations. *J.Macromol.Sci., Part A, Pure & Appl. Chem.* 50(5), 449-465 (2013).

N.Zavradashvili, T.Memenishvili, N. Kupatadze, L.Baldi, X.Shen, D.Tugushi, C.Wandrey, R.Katsarava. Cell compatible arginine containing cationic polymer: one-pot synthesis and preliminary biological assessment, *Springer Book Series-Advances in experimental medicine and biology: Infectious Diseases and Nanomedicine*, 2014, 59-73.

T. Memanishvili, N. Zavradashvili, N. Kupatadze, D. Tugushi, M. Gverdtsiteli, V.P. Torchilin, C. Wandrey, L. Baldi, S. S. Manoli, R. Katsarava. Arginine-based biodegradable ether-ester polymers of low cytotoxicity as potential gene carriers. *Biomacromolecules*, 15, 2839-2848 (2014).

A.Díaz, R.Katsarava, J.Puiggalí, Synthesis, properties and applications of biodegradable polymers derived from diols and dicarboxylic acids: from polyesters to poly(ester amide)s. *Int. J. Mol. Sci.* 15, 7064-7123 (2014).

A.Díaz, L. J. del Valle, D. Tugushi, R. Katsarava, J. Puiggalí, New poly(ester urea) derived from L-leucine: electrospun scaffolds loaded with antibacterial drugs and enzymes. *Materials Science and Engineering C (in press).*

M. Planellas, M. M. Pérez-Madrigal, L. J. del Valle, S. Kobauri, R. Katsarava, C. Alemán, J. Puiggalí, Microfibres of conducting polythiophene and biodegradable poly(ester urea) for scaffolds, *Polymer Chemistry (in press).*

L. El Haddad, N. Ben Abdallah, P.-L. Plante, J. Dumaresq, R. Katsarava, S. Labrie, J. Corbeil, D.St-Gelais, S. Moineau, Improving the safety of a *Staphylococcus aureus* polyvalent phage by its production on a *Staphylococcus xylosus* strain, *PLoS ONE (in press).*

In: Biochemistry Research Summaries. Volume 1
Editors: James T. Walker and Catherine M. Jones
ISBN: 978-1-53610-895-8

Chapter 12

GISELA KRISTOFF

Affiliation:
IQUIBICEN, UBA-CONICET; Departamento de Química Biológica, Facultad de Ciencias Exactas y Naturales, Universidad de Buenos Aires.

Contact Point:
Juramento 1155 8 A, Buenos Aires, Argentina

Date of Birth:
02/06/71

Education:
Biochemistry, Facultad de Farmacia y Bioquímica, Universidad de Buenos Aires; Doctor (PhD), Facultad de Ciencias Exactas y Naturales, Universidad de Buenos Aires.

Research and Professional Experience:
Gisela Kristoff is an investigator in the area of toxicology. She has 13 papers and one book published in the area. Also she has presented 37 works in national and international congress. She learns in the University of Buenos Aires, actually in the area of toxicology. She was the director of several thesis and actually she has two in course.

Professional Appointments:
Toxicology, biochemistry, aquatic toxicology.

Honors:
PhD thesis: "Estudio comparativo de biomarcadores en los invertebrados acuáticos *Biomphalaria glabrata* y *Lumbriculus variegatus* expuestos a pesticidas de relevancia ambiental" has obtained a Price: the best thesis in toxicological area, Asociación Química Argentina, 2012.

Publications Last 3 Years:
-Kristoff, G., 2013. ¿Son las carboxilesterasas biomarcadores más sensibles que las colinesterasas? Conociendo un poco más a los biomarcadores para insecticidas

anticolinesterásicos en invertebrados acuáticos. Editorial Académica Española, Saabrucken, Deutschland, Alemania. ISBN: 978-3-659-06554-5

- Rivadeneira, P. R., Agrelo, M., Otero, S., Kristoff G. Different effects of subchronic exposure to low concentrations of the organophosphate insecticide chlorpyrifos in a freshwater gastropod, 2013. Ecotoxicology and Environmental Safety, 90, 82-88.

- Bianco, K., Yusseppone, M. S., Otero, S., Luquet, C., Ríos de Molina, M. C., Kristoff, G., 2013. Cholinesterases and neurotoxicity as highly sensitive biomarkers for an organophosphate insecticide in a freshwater gastropod (*Chilina gibbosa*) with low sensitivity carboxylesterases. Aquatic Toxicology, 144-145, 26-35, 2013.

-Bianco, K., Otero, S., Balazote Oliver, A., Nahabedian, D., Kristoff, G., 2014. Resistance in cholinesterase activity after an acute and subchronic exposure to azinphos-methyl in the freshwater gastropod *Biomphalaria straminea*. Ecotoxicology and Environmental Safety, 109, 85-92.

In: Biochemistry Research Summaries. Volume 1
Editors: James T. Walker and Catherine M. Jones
ISBN: 978-1-53610-895-8

Chapter 13

ELMAR W. KUHN, MD

Affiliation:
Department of Cardiothoracic Surgery, Heart Center, University of Cologne, Germany

Contact Point:
Kerpener Strasse 62, 50924 Cologne, Germany

Date of Birth:
5th October, 1981

Education:
2001 to 2007: University of Cologne, Medical School

2005 / 2006: Rotations and electives in Presbyterian Hospital, Columbia University, New York, USA and Massachusetts General Hospital, and Harvard University, Boston, USA

2007: Medical Doctor

Research and Professional Experience:
2009: German Cochrane Center Certificate, Freiburg, Germany

2010: GCP Training "Investigator", Clinical Trials Center Cologne

2010 to 2011: Rotation to Institute for Experimental Medicine, University of Cologne; Project: "Rosuvastatin Reloading before Cardiac Surgery with Cardiopulmonary Bypass"

2011 to 2012: Rotation Clinical Trials Center Cologne

2011: GCP Training "Principal Investigator", Clinical Trials Center Cologne

2011: Cardiac Surgeon Curriculum Module I: Aortic valve, Brussels, Belgium

2011: Cardiac Surgeon Curriculum Module II: Mitral valve, Brussels, Belgium

2012: Cardiac Surgeon Curriculum Module III: Future Techniques, Brussels, Belgium

2013: Master Course, Aortic and Mitral Replacement, Nyon, Switzerland

Investigator in clinical trials/registries: StaRT-CABG Trial (investigator initiated trial, www.start-cabg.de), Sapien 3 (Edwards Lifesciences), JUPITER (JenaValve), Perigon (Medtronic), INGEVITY (Boston Scientific)

Professional Appointments:

Residency, Department of Cardiothoracic Surgery, University of Cologne, Head: Univ.-Prof. Dr. T. Wahlers

Publications Last 3 Years:

1. Kuhn* EW, Liakopoulos* OJ, Stange S, Deppe A, Slottosch I, Choi CY, Wahlers T. Preoperative statin therapy in cardiac surgery: a meta-analysis of 90 000 patients. European journal of cardio-thoracic surgery: official journal of the European Association for Cardio-thoracic Surgery 2013:1-10.

2. Kuhn EW, Liakopoulos OJ, Deppe AC, Slottosch I, Neef K, Sterner-Kock A, Madershahian N, Choi YH, Wahlers T. Rosuvastatin Reloading before Cardiac Surgery with Cardiopulmonary Bypass. European surgical research 2013;50:1-13.

3. Deppe AC, Kuhn E, Scherner M, Slottosch I, Liakopoulos O, Langebartels G, Choi YH, Wahlers T. Coagulation Disorders Do Not Increase the Risk for Bleeding during Percutaneous Dilatational Tracheotomy. The Thoracic and cardiovascular surgeon 2013.

In: Biochemistry Research Summaries. Volume 1
Editors: James T. Walker and Catherine M. Jones
ISBN: 978-1-53610-895-8

Chapter 14

IVANA LONCAREVIC

Affiliation:

Faculty of Technology, University of Novi Sad, Department of Engineering Carb Foods

Contact Point:

Bulevar cara Lazara 1, 21000 Novi Sad, Serbia

Date of Birth:

7th May 1982

Education:

PhD in Food Engineering

Research and Professional Experience:

Research field based on the influence of edible fats and oils, as well as emulsifiers on crystallization characteristics, physical properties and shelf life of confectionery products.

Professional Appointments:

Research Associate at Department of Engineering Carb Foods.

Honors:

Participation on the European workshop of best PhD works in the field of science of food engineering and food technology, The 8th European Conference on Food Engineering and Technology, held in April 1st – 2nd 2014., in Quakenbrück, Germany.

Publications Last 3 Years:

I. Lončarević, B. Pajin, R. Omorjan, A. Torbica, D. Zarić, J. Maksimović, J. Švarc Gajić (2013): The influence of lecithin from different sources on crystallization and physical properties of non trans fat, Journal of Texture Studies, 44, 450-458.

A. Torbica, B. Pajin, R. Omorjan, I. Lončarević, J. Tomić (2014): Physical properties of chocolate with addition of Cocoa Butter Equivalent of moderate hardness, Journal of the American Oil Chemists Society, 91, 39-48.

I. Lončarević, B. Pajin, Lj. Dokić, Z. Šereš, A. Fišteš, D. Šoronja Simović, V. Krstonošić (2014): Rheological and textural properties of cocoa spread cream with sunflower lecithin, Acta Tehnica Corviniensis – Bulletin of Engineering, (7) 47-50.

PART II

RESEARCH SUMMARIES

In: Biochemistry Research Summaries. Volume 1
Editors: James T. Walker and Catherine M. Jones
ISBN: 978-1-53610-895-8

Chapter 15

ORGANIC CHEMISTRY, BIOCHEMISTRY, BIOTECHNOLOGY AND RENEWABLE RESOURCES. RESEARCH AND DEVELOPMENT[#]

VOLUME 1
TODAY AND TOMORROW

Gennady E. Zaikov, Frank Pudel and Grzegorz Spychalski (Editors)
N. Emmanuel Institute of Biochemical Physics
Russian Academy of Sciences, Moscow, Russia

RESEARCH SUMMARY

We included in this volume very important (and last) information in the field of organic chemistry, biochemistry, biotechnology, material science (synthesis, properties, applications) and renewable resources.

We did special emphases on the problems of microbiological corrosion of metals and protection against it, biodamaging and protection of paint materials, biodamages and protection of paper, biodegradation and protection of synthetic polymer materials, biodamages and protection of wood, coordination of specific retention volumes of organic compounds with their physical–chemical properties by means of polyparametric equations, renewable resources and plant biotechnology and hemp essential oil - production, characterization, application, enriched linen textiles as a material for "wellness" products.

[#] This chapter was previously published as a book: *Organic Chemistry, Biochemistry, Biotechnology and Renewable Resources. Research and Development. Volume 1 - Today and Tomorrow*, edited by Gennady E. Zaikov, Frank Pudel and Grzegorz Spychalski, New York: Nova Publishers Inc., 2013.

Parts of the volume is devoted to use of plant biotechnology methods for flax breeding in Latvia, exploring the potential of oilseeds as a sustainable source of oil and protein for aquaculture feed, potential and future prospects of dyeing plants: Madder *Rubia tinctorium* L., Annato *Bixa orellana* L, Dyer's broom *Genista tinctoria* L., Woad *Isatis tinctoria.* L., genetic improvement of oat hulls and their potential for use as a renewable resource, application of the ultrasound treatment of flax fibers to produce yarns by open-end and wet flax spinning, preparation of rapeseed protein products for aquaculture feeding, sustainable biofuels - an international overview of existing approaches, Polish cultivar of white mulberry (*Morus alba* L.) and new directions of its utilization, evaluation of different metal oxide nano particle finishes, for self-cleaning of cotton fabric, biomass logistics: the key challenge of minimizing supply costs, ecological and economic optimisation of biogas plants.

Only fascination of science is able to overcome natural disgust to the efforts of mind – all of us (Editors and contributors of this volume) were permanently deliberated these Isaak Newton's words when we have move along the reads of discoveries and understanding of the new idea in Organic Chemistry, Biochemistry, Biodegradation and Creation of New Materials on the base of Renewable Resources. These were the roads following the Creator's reason (please look for the epigraph of this volume above), the roads of acquisitions and losses, successful probes and humiliating errors, inspiration and despair, sacred delight and hopeless dejection. We were very happy to cooperate with contributors of this book.

In: Biochemistry Research Summaries. Volume 1
Editors: James T. Walker and Catherine M. Jones
ISBN: 978-1-53610-895-8

Chapter 16

RESVERATROL FROM VINE TO WINE[#]

Marielle Adrian*,[1] Mirko De Rosso,[2] Luigi Bavaresco,[2] Benoît Poinssot[1] and Marie-Claire Héloir[1]

[1] UMR 1347 Agroécologie AgroSup/INRA/uB
Pôle Interactions Plantes-Microorganismes-ERL CNRS,
Dijon cedex, France
[2]Centro di Ricerca per la Viticoltura,
Consiglio per la Ricerca e la Sperimentazione in Agricoltura,
CRA – VIT, Conegliano, Italy

RESEARCH SUMMARY

The stilbene compound resveratrol and its derivatives are produced by various plant families and possess a wide spectrum of interesting medical properties. In *Vitaceae,* these compounds are phytoalexins *ie* antifungal metabolites synthesized and accumulated in response to biotic or abiotic stresses.

Therefore, even if their level in wines is influenced by winemaking process and fermentative micro-organisms, it largely depends on their initial level in berries. Resveratrol is synthesized *via* the phenylalanine / polymalonate pathway and can undergo glycosylation, methylation or polymerization, leading to a set of derivatives. Due to their distinct chemical structures, these molecules possess different activity and efficiency regarding their biological properties.

Vitis species and *Vitis vinifera* varieties do not have the same genetic ability to produce these compounds and there is a positive correlation between their ability to accumulate them and their level of resistance to pathogens. In this chapter, we propose to list different biotic and abiotic factors able to induce stilbene accumulation in grapevine organs.

This chapter was previously published in *Resveratrol: Sources, Production and Health Benefits,* edited by Dominique Delmas, New York: Nova Publishers Inc., 2013.

* Corresponding author: Marielle Adrian. UMR 1347 Agroécologie AgroSup/INRA/uB, Pôle Interactions Plantes-Microorganismes-ERL CNRS 6300, 17 rue Sully, BP 86510, 21065 Dijon cedex, France.

In: Biochemistry Research Summaries. Volume 1
Editors: James T. Walker and Catherine M. Jones
ISBN: 978-1-53610-895-8

Chapter 17

PRODUCTION OF GRAPEVINE CELL BIOMASS (*VITIS VINIFERA L.* GAMAY) AND RESVERATROL IN CUSTOM AND COMMERCIAL BIOREACTORS USING CYCLODEXTRINS AND METHYL-JASMONATE ELICITORS[#]

Juan Carlos Vera-Urbina[1], Susana Sellés-Marchart[1], María José Martínez-Esteso[1], María Angeles Pedreño[2] and Roque Bru-Martinez[1,*]

[1]Department of Agrochemistry and Biochemistry, Faculty of Science, University of Alicante, Alicante, Spain
[2]Department of Plant Physiology, Faculty of Biology, University of Murcia, Murcia, Spain

RESEARCH SUMMARY

Plant cell liquid culture may become an alternative to produce valuable secondary metabolites instead of extracting from plant raw material. Besides the optimization of the culture medium and culture conditions of highly productive cell lines, the culture scale-up in a suitable bioreactor is a compulsory step for bioprocess development. We have designed an airlift-type bioreactor (ALB) of low cost featured for aseptically whole medium extraction and replacement. Medium exchange enables the modification of the culture conditions and the recovery of metabolites secreted to the medium as advantageous features. The suitability of ALB for plant cell culture has been tested by comparing its performance with the commercial stirred and aerated tank bioreactor (SATB)-type Biostat B to grow up grapevine cells. The comparison has been realized under a randomized block experimental design,

[#] This chapter was previously published in *Resveratrol: Sources, Production and Health Benefits,* edited by Dominique Delmas, New York: Nova Publishers Inc., 2013.

[*] E-mail: Roque.Bru@ua.es.

taking the bioreactor type and the sucrose concentration as factors. Biomass yield (g biomass. substrate^{-1}), productivity (g biomass. g substrate^{-1}.day^{-1}), maximum biomass production, final biomass concentration, maximum growth rate, sucrose consumption rate and glucose usage were determined at three different levels of sucrose concentration, low, medium and high (10, 20, 30 g.L^{-1} respectively) in both bioreactors in two replicates. The cell growth rate during the exponential phase (μ) was not significantly affected by either the bioreactor type or the sucrose concentration, but in the SATB the exponential phase extended for longer times than in ALB, thus leading to obtain more final biomass and higher productivity variables. Lower oxygen availability could be a causal factor for the less efficient use of the carbon source in ALB, but this limitation might be solved by a redesign of an improved ALB prototype. To achieve extracellular resveratrol production, culture media were supplemented with elicitors and incubated with cells in the bioreactors for several days. Upon elicitation with methylated cyclodextrin, the medium in ALB reached a final concentration of 0.58 mg.mL^{-1} and a specific production of 2.4 mg. g biomass^{-1} while in STAB was 1.01 mg.mL^{-1} and 3.1 mg. g biomass^{-1} respectively. When methylated cyclodextrin were used in combination with methyl jasmonate the resveratrol production largely increased and was similar in both bioreactors, final concentration above 4.50 mg.mL^{-1} and specific production above 13.0 mg. g biomass^{-1}. Although the growth of grapevine cells in ALB was not as effective as in STAB, the design of ALB bioreactor showed a great potential for the production of resveratrol.

In: Biochemistry Research Summaries. Volume 1
Editors: James T. Walker and Catherine M. Jones
ISBN: 978-1-53610-895-8

Chapter 18

DEVELOPMENT OF VOLTAMMETRIC TECHNIQUES AND SENSORS FOR THE DETERMINATION OF RESVERATROL[#]

Adrian Marcelo Granero,[*] Fernando Javier Arévalo, Héctor Fernández and María Alicia Zon[±]
Departamento de Química, Facultad de Ciencias Exactas, Físico-Químicas y Naturales, Universidad Nacional de Río Cuarto, Río Cuarto, Argentina

RESEARCH SUMMARY

Resveratrol is a polyphenolic compound of the viniferins family. Resveratrol exists in nature in both the *trans*- and *cis*-stereoisomeric forms. However, the *trans*-resveratrol or *trans*-3,4,5'-trihydroxystilbene (t-Res) is the most biologically active form. The t-Res was discovered by Michio Takaoka more than 70 years ago, in the resin of *Veratrum grandiflorum*. The t-Res occurs naturally in more than 80 plant species, such as berries, peanuts, groundnuts, tea, grapevines and red wines. However, the most abundant natural sources of t-Res are *Vitis vinifera*, *labrusca* and *muscadine* grapes, which are used to make wines. Besides its natural occurrence, t-Res is now available in tablets, being recommended as a dietary supplement.

Numerous previous studies have investigated many of its beneficial effects, such as an antioxidant, phytoestrogen, lifespan prolongation, anti-inflammatory, antiplatelet, antiatherosclerosis. In addition, t-Res is identified as a preventive agent against several pathologies: cancer, viral infection, neurodegenerative processes and cardiovascular diseases. Moreover, because of its high concentration in grape skin, red wines contain significant amount of t-Res, which it is thought to be responsible for the so called the "French paradox",

[#] This chapter was previously published in *Resveratrol: Sources, Production and Health Benefits,* edited by Dominique Delmas, New York: Nova Publishers Inc., 2013.
[*] Corresponding author: Adrian Marcelo Granero. E-mail: agranero@exa.unrc.edu.ar.
[±] Departamento de Química, Facultad de Ciencias Exactas, Físico-Químicas y Naturales, Universidad Nacional de Río Cuarto, Agencia Postal N° 3 (5800) - Río Cuarto, Argentina.

i.e., low mortality due to coronary heart diseases as a result of a moderate consumption of red wines.

The growing interest in the determination of t-Res in food is directly related to the need of developing sensitive methods for its analysis. Thus, many methods have been developed for the determination of t-Res in food, mainly different chromatographic methods, particularly HPLC using different detectors (UV, fluorescence, capillary electrophoresis, electrochemical, mass spectroscopy and diode array detection).

Because of the t-Res electrochemical properties, it can also be quantified using different electroanalytical methods. The more attractive features of voltammetric methods are the high sensitivity, low cost, simplicity and relatively short measurement time. Voltammetric techniques in general and stripping techniques in particular, have been employed in food analysis, including wine among the considered matrices.

Another interesting proposal is the use of biosensors. According to the IUPAC definition, a biosensor is a device that uses specific biochemical reactions mediated by isolated enzymes, immunosystems, tissues, organelles or whole cells to detect chemical compounds usually by electrical, thermal or optical signals. The biosensors have been proposed as an effective analytical tool for the determination of polyphenolic compounds, exhibiting advantages such as the minimal preparation of the sample, selectivity, sensitivity, reproducibility, rapid time of response and simple use for continuous on-site analysis.

In: Biochemistry Research Summaries. Volume 1
Editors: James T. Walker and Catherine M. Jones
ISBN: 978-1-53610-895-8

Chapter 19

ADSORPTION PROPERTIES OF SYNTHETIC DETERMINATIONS OF RESVERATROL[#]

Konstantin E. Polunin and Irina A. Polunina*
Frumkin Institute of Physical Chemistry and Electrochemistry,
Russian Academy of Sciences, Moscow, Russia

RESEARCH SUMMARY

Much research all over the world is devoted to interactions of biomolecules with the surface of a solid phase because of the great practical importance of this problem. These data become of current interest in connection with the development of nanobiotechnology seeking to create novel composite materials of the "inorganic nanounit–biomolecule" type. However, investigations of the behavior of actual medicines at interfaces are very scarce.

The goal of this study was to examine the adsorption and physicochemical properties of synthetic resveratrol (*trans*-3,5,4'-trihydroxystilbene) and other *trans*-stilbenoids in a nonpolar organic media at the solid/liquid interface.

Normal-phase chromatography (TLC, HPLC) and UV spectroscopy are used to study the dynamic adsorption of stilbenoids at silica gel in the range of low concentrations in organic solvents. Equilibrium adsorption isotherms of synthetic *trans*-hydroxystilbenes at nanodispersed silicon, titanium, and aluminum oxides were measured under static conditions.

It was found that adsorption of hydroxystilbenes depends on the number and geometric positions of OH groups in the molecular structure as well as on the surface chemistry of the oxides and the organic solvent composition. The adsorption of resveratrol is best at titania and worst at silica. Compared to other hydroxystilbenes, the adsorption of 3,5,4',5'-tetrahydroxystilbene (piceatannol) is the best at the oxides.. It was found that the adsorption behavior of hydroxystilbenes can be modeled based on the data on the adsorption of phenols in this system. The adsorption behavior of resveratrol is best modeled with phloroglucinol.

[#] This chapter was previously published in *Resveratrol: Sources, Production and Health Benefits,* edited by Dominique Delmas, New York: Nova Publishers Inc., 2013.

* E-mail: polunina@phyche.ac.ru.

In: Biochemistry Research Summaries. Volume 1
Editors: James T. Walker and Catherine M. Jones
ISBN: 978-1-53610-895-8

Chapter 20

A Fast and High-Throughput MEPS/UHPLC-PDA-Based Methodology for Separation and Identification of Resveratrol in Wine[#]

João Gonçalves and José S. Câmara*
Centro de Química da Madeira, Centro de Ciências Exactas e da Engenharia da Universidade da Madeira, Campus Universitário da Penteada, Funchal, Portugal

Research Summary

Grapes and grape-derived products such as red wines are an interesting source of polyphenols and represent an important dietary component. The natural stilbene, resveratrol (3,4',5-trihydroxystilbene), is a nonflavonoid that appears to be one of the most biologically active polyphenols. In grapevines (*Vitis vinifera L.*), resveratrol was synthesized in response to biotic and abiotic stress, such as fungal infection, ultra violet light exposure, ozone stress, anoxic treatment, and wounding. Resveratrol exists as *cis-* (*Z*) and *trans-* (*E*) isomers, both of which are found in wines, however the *trans-* isomer occurs predominantly and the presence of the *cis-* isomer, in certain wines, is attributed to photoisomeric conversion, enzyme action during fermentation, or release from viniferins. The *trans*-resveratrol has been shown to be the more biologically active form of resveratrol. Their very simple chemical structure may interact with a variety of cellular and molecular targets.

Several studies have demonstrated that resveratrol provides health benefits such as protection against coronary heart diseases, atherosclerosis, cancer, and reduction of the effects of some neurologic diseases, such as, Alzheimer, Parkinson, among others. The methodology used to analyse this phenolic compound in wine generally includes extractions with solvents. Conventional liquid-liquid extraction (LLE) and solid-phase extraction (SPE) are the most

[#] This chapter was previously published in *Resveratrol: Sources, Production and Health Benefits*, edited by Dominique Delmas, New York: Nova Publishers Inc., 2013.

* Tel: +351 291705112; Fax: +351 291705149 E-mail address: jsc@uma.pt.

common. In last years, miniaturized analytical extraction techniques, such as solid-phase microextraction (SPME) and stir bar sorptive extraction (SBSE), both in combination with high performance liquid chromatography (HPLC), had gained attention due to its many special features over classical approaches. Another miniaturized technique is the microextraction by packed sorbent (MEPS) appears to overcome some of drawbacks present in conventional methods, retaining the principles of SPE while offering several advantages that include speed, less solvent consumption and improve the recovery efficiency.

Hence, an ultrafast and improved analytical methodology based on MEPS combined with ultra-high performance liquid chromatography (UHPLC) equipped with a photodiode array (PDA) detector, was developed and validated for determination of *(E)*-resveratrol in wines. The method was fully validated in terms of selectivity, linearity, detection (LOD) and quantification (LOQ) limits, extraction yield, accuracy and inter/intra-day precision. Validation experiments revealed very good recovery rate of 95±5.8% RSD, good linearity with r^2*-values* > 0.999 within the established concentration range, excellent repeatability (0.52%) and reproducibility (1.67%) values (expressed as RSD), thus demonstrating the robustness and accuracy of the $MEPS_{C8}$/UHPLC-PDA method. The limit of detection (LOD) of the method was 0.21 μg mL^{-1} whereas the limit of quantification (LOQ) was 0.68 μg mL^{-1}. The validated methodology was applied to thirty commercial wines (24 red wines and 6 white wines) from different grape varieties, vintage and regions. On the basis of the analytical validation, the $MEPS_{C8}$/UHPLC–PDA methodology show to be an improved, sensitive, and ultrafast approach for determination of (*E*)-resveratrol in wines with high resolving power within 6 min.

In: Biochemistry Research Summaries. Volume 1
Editors: James T. Walker and Catherine M. Jones

ISBN: 978-1-53610-895-8

Chapter 21

ANTIOXIDANT PROPERTIES OF RESVERATROL AND HEALTH PROTECTION IN PATHOPHYSIOLOGICAL STATES[#]

Ramón Rodrigo*[*], David Gil-Becerra and Andrés Miranda-Merchak
Molecular and Clinical Pharmacology Program,
Institute of Biomedical Sciences, Faculty of Medicine,
University of Chile, Chile

RESEARCH SUMMARY

Oxidative stress plays a key role in the development and progression of several diseases such as cancer, cardiovascular disease, diabetes, osteoporosis and neurodegenerative diseases, among others. Recent studies suggest that antioxidants agents may provide an effective approach to quench oxidative stress in tissues. It has been suggested that polyphenols are involved in protection against the mechanisms leading to structural and functional organ damage. Resveratrol, a natural phytoalexin polyphenolic compound found in the skin of grapes, as well as in red wine, may play an important role in the prevention of human diseases through its pleiotropic actions. The overall protective effect of resveratrol is due to its large array of biological actions, such as free radical-scavenging, metal chelation, enzyme regulation, cell signalling pathways and gene expression modulations.

There have been reported no adverse effects derived from resveratrol intakes currently associated with the normal diet. However, supplements for health protection should be cautiously used as no level definition has been given to make sure the dose is safe. The therapeutic potential of resveratrol, in cancer chemoprevention and cardioprotection, has stimulated many studies respect to the possible mechanisms of action. Future large scale and

---

[#] This chapter was previously published in *Resveratrol: Sources, Production and Health Benefits,* edited by Dominique Delmas, New York: Nova Publishers Inc., 2013.

[*] Tel: 56-2-9786126. Fax: 56-2-9786126. E-mail: rrodrigo@med.uchile.cl.

randomized clinical trials should be conducted to fully establish the therapeutic use of resveratrol against human disease. This review focuses on antioxidant properties of resveratrol and its importance in human health and disease.

In: Biochemistry Research Summaries. Volume 1
Editors: James T. Walker and Catherine M. Jones
ISBN: 978-1-53610-895-8

Chapter 22

RESVERATROL AND OXIDATIVE STRESS OF AGEING[#]

Qingzhao Ruan[1] and Runsheng Ruan[2*]
[1] Royal Free and University College Medical School,
University College London, Gower Street, London
[2] Biotherapy Centre, GIBH Affiliated Fuda Hospital,
Chinese Academy of Sciences,
Jude Zhong Lu, Chi Gang, Guangzhou, China

RESEARCH SUMMARY

Age and ageing are typically understood as a chronological register of one's process of maturation and development. The concept of biological age drew emphasis away from ageing being a mere process of time-keeping and instead focuses on the functional acuity of our physiology, which is in turned governed by various biochemical pathways on the intracellular level. Products accumulating with normal cycling of these pathways are identified as markers of senescence, and some are known to interfere with the pathways themselves.

Several agents have been popularized as determinants of outcomes of the above pathways, and reactive oxygen species (ROS), backed by free radical theory, have been placed under intense scrutiny. ROS is thus not only a marker of cell age, but also a target for elimination in anti-ageing formulae. Numerous studies exploring the mechanisms of ROS-induced senescence yielded several fairly concordant pathways of damage. First of all, oxidative injury to intracellular proteins promotes inflammatory changes quantifiable by protein carbonyl content (PCC) assay. Non-enzymatic radical-driven lipid peroxidation measured using 8-iso-$PGF_{2\alpha}$ serves as another mean of identifying oxidative damage. Finally, DNA single or double-stranded breaks following oxidative insult produces markers following the radical attack. The product, 8OHdG, is well correlated with the concentration of 8-iso-$PGF_{2\alpha}$ in our previous study. Going a step further, ROS insult within mitochondria dissipates the integrity of the mitochondrial inner membrane. We have shown that the membrane

[#] This chapter was previously published in *Resveratrol: Sources, Production and Health Benefits,* edited by Dominique Delmas, New York: Nova Publishers Inc., 2013.

* E-mail : rsruan@singnet.com.sg.

becomes progressively deploarized as age progresses, and this is another crucial aspect worthy of note in the context of ageing.

Immunosenescence, understood as deteriorating efficacy in one's immune system with age, explains the increasing susceptibility of the old to common infections. Deranged T cell function and its receptor expression amongst others are influenced by mechanisms of oxidative insult. It is the potential reversal of all these biological changes that brought Resveratrol into the limelight.

Resveratrol (RSV) began gaining significance in the scientific community when it was suggested that it possesses cardioprotective effects made apparent in the French paradox. RSV, a naturally occuring phytoalexin, is synthesised by grapes in response to fungal infection, and is therefore found in red wine as well as peanuts and berries. Through mechanisms not entirely understood, it has been shown recently to extend both mean and maximum life span in organisms such as yeast and metazoans. A whole host of likely pathways have been proposed, including the prevention of lipid peroxidation, selective COX inhibition, stimulation of Sir2 which increases DNA stability and the effective scavenging of mitochondrial oxygen species. Overall, it have been reported to mimic the effects of caloric restriction, which is currently one of the few recognized measures of anti-ageing. We have shown that in age-associated immune changes, that low dose administration of RSV promotes a T-cell profile and cytokine environment that resembles that of the young. This is highly relevant in the demographic context of today where pathologies, including influenza and malignancies, brought on by imbalances in the senescent immune system constitute a major cause of mortality in the aged.

In: Biochemistry Research Summaries. Volume 1
Editors: James T. Walker and Catherine M. Jones
ISBN: 978-1-53610-895-8

Chapter 23

ANTIOXIDANT PROPERTIES OF RESVERATROL AND ITS PROMISING EFFECTS IN CARDIOVASCULAR DISEASE[#]

Dominique Bonnefont-Rousselot*
Service de Biochimie Métabolique,
Groupe Hospitalier Pitié-Salpêtrière-Charles Foix, Paris Cedex, France
Département de Biologie Expérimentale, Métabolique et Clinique (EA 4466),
Faculté des Sciences Pharmaceutiques et Biologiques,
Université Paris Descartes, Paris, France

RESEARCH SUMMARY

Resveratrol (3,4',5-trans-trihydroxystilbene) (RES) is a phytoestrogen found in biological materials such as grapes, grape juice, red wine, plums and peanuts. Several studies have shown its beneficial effects on neurological, hepatic and cardiovascular systems. It has been suggested that the antioxidant properties of resveratrol could be partly responsible for the protective effect against cardiovascular diseases, observed with consumption of moderate amounts of red wine. Resveratrol has been reported to be a good antioxidant against lipid peroxidation in cells, membranes, and lipid models (micelles, low density lipoproteins). RES structural analogues like piceatannol (i.e., 3,3',4',5-trans-tetrahydroxystilbene) (PIC) have also been studied. PIC, which is a metabolite of RES via cyt.P450 1A2, has been reported to exhibit a higher antioxidant effect than RES, partly due to the presence of an additional hydroxyl group (competition with the fatty acid to scavenge lipid peroxyl radicals). The ability of RES to scavenge peroxyl radicals involved in lipid peroxidation could contribute to the beneficial effect of RES in cardioprotection. Besides, oxidative stress, that contributes to the development and progress of cardiovascular diseases, results in endothelial dysfunction

---

[#] This chapter was previously published in *Resveratrol: Sources, Production and Health Benefits*, edited by Dominique Delmas, New York: Nova Publishers Inc., 2013.

* Tel : (33) 1 42 16 20 58 Fax : (33) 1 42 16 20 33 E mail: dominique.rousselot@psl.aphp.fr.

whose mechanism includes both decreased availability of nitric oxide (•NO), a potent vasodilator, and altered expression of the endothelial NO-synthase (eNOS). eNOS activity is regulated by endogenous inhibitors, particularly asymmetric dimethylarginine (ADMA), which is metabolized by the dimethylarginine dimethylaminohydrolase (DDAH). Thus, a reduced DDAH activity has been reported in hypercholesterolemia and diabetes mellitus, both pathological conditions characterized by increased levels of reactive oxygen species (ROS). Recent studies confirmed the redox-sensitive regulation of DDAH, and a significant decrease of DDAH activity and expression has been reported in bovine aortic endothelial cells cultured under conditions of high glucose concentrations in order to mimic the pathophysiological stimulus of the diabetic state. Under these experimental conditions, RES and PIC restored DDAH activity in a dose-dependent manner, with a more efficient effect of PIC. In addition, RES induces vasodilation by counterbalancing the effect of the vasoconstrictor endothelin-1, and exhibits anti-inflammatory and antithrombotic effects, so that it could be considered as a potential adjuvant to promote cardiovascular health.

In: Biochemistry Research Summaries. Volume 1
Editors: James T. Walker and Catherine M. Jones
ISBN: 978-1-53610-895-8

Chapter 24

MOLECULAR MECHANISMS ON HOW RESVERATROL ENHANCES ENDOTHELIAL NO PRODUCTION[#]

Huige Li[*] and Ulrich Förstermann
Department of Pharmacology, University Medical Center,
Johannes Gutenberg University, Mainz, Germany

RESEARCH SUMMARY

Nitric oxide (NO) derived from the endothelial NO synthase (eNOS) is a potent vasodilator with multiple anti-atherosclerotic properties. Enhancement of endothelial NO production may have therapeutic potential against cardiovascular disease. Resveratrol increases endothelial NO production through multiple mechanisms: (i) resveratrol stimulates eNOS gene transcription; (ii) resveratrol enhances eNOS mRNA stability; (iii) resveratrol increases eNOS enzymatic activity by enhancing eNOS phosphorylation; (iv) resveratrol augments eNOS activity through sirtuin 1-mediated deacetylation; (v) resveratrol reduces the levels of asymmetric dimethylarginine (ADMA), which is an endogenous eNOS inhibitor; (vi) resveratrol improves eNOS functionality by preventing eNOS uncoupling; (vii) resveratrol inhibits superoxide-mediated NO inactivation. Signal cascades of the abovementioned mechanisms are summarized in this chapter.

[#] This chapter was previously published in *Resveratrol: Sources, Production and Health Benefits,* edited by Dominique Delmas, New York: Nova Publishers Inc., 2013.

[*] E mail: huigeli@uni-mainz.de.

In: Biochemistry Research Summaries. Volume 1
Editors: James T. Walker and Catherine M. Jones
ISBN: 978-1-53610-895-8

Chapter 25

Basis for Resveratrol's Potential for Prevention and Treatment of Obesity and Osteoporosis[#]

Clifton A. Baile[1,2,*], Mary Anne Della-Fera[1] and Srujana Rayalam[1]
[1]Department of Animal and Dairy Science,
University of Georgia, Athens, Georgia, US
[2]Department of Foods and Nutrition,
University of Georgia, Athens, Georgia, US

Research Summary

Higher levels of adiposity are associated with increased risk for development of numerous adverse health conditions including type-two diabetes, bone disorders and cancer. Phytochemicals and combinations of phytochemicals are potential supplements for humans to suppress differentiation of preadipocytes, stimulate lipolysis, and induce apoptosis of existing adipocytes, thereby providing a preventive and treatment for adiposity. Resveratrol decreases adipogenesis and viability in maturing preadipocytes which has been shown to be through down-regulating adipocyte specific transcription factors and enzymes and gene products that change mitochondrial function. Furthermore, resveratrol increases fat mobilization and synthesis in mature adipocytes. Potentially of special importance is that resveratrol combined with other natural products produces synergistic activities from the modulation of multiple biotargets throughout the life cycle of adipocytes. It was reported that mice treated with resveratrol alone reduced weight gain caused by a high fat diet. Because mesenchymal stem cells are precursors for both adipocytes and osteoblasts and the differentiation to adipocytes dominates over the differentiation to osteoblasts in bone marrow of aging humans, there is an

---

[#] This chapter was previously published in *Resveratrol: Sources, Production and Health Benefits,* edited by Dominique Delmas, New York: Nova Publishers Inc., 2013.

[*] Tel: 706-542-4094; Fax: 706-542-7925; E-mail: cbaile@uga.edu.

increased propensity for osteoporosis and bone fracutres. Resveratrol, and other phytochemicals, act on several biotargets in adipocytes and osteoblasts leading to a decrease in adipocyte number and size and an increase in osteogenesis. In a post-menopausal animal model, aged ovariectomized rats, dietary supplementation using a combination of resveratrol with vitamin D, quercetin and genistein decreased weight gain and inhibited bone loss. Combining several phytochemicals, including especially resveratrol, is likely to lead to combinations that are efficacious for the prevention and treatment of obesity and osteoporosis through targeted decrease in adipogenesis, enhanced apoptosis and lipolysis in adipocytes and increased osteogenesis in osteoblasts.

In: Biochemistry Research Summaries. Volume 1
Editors: James T. Walker and Catherine M. Jones
ISBN: 978-1-53610-895-8

Chapter 26

RESVERATROL AND ENERGY METABOLISM[#]

Marc Pignitter and Veronika Somoza*
Department of Nutritional and Physiological Chemistry,
University of Vienna, Vienna, Austria

RESEARCH SUMMARY

Resveratrol, a natural polyphenolic stilbene derivate, has been shown to exert multiple effects on energy metabolism by modulating different proteins involved in energy homeostasis. Specifically, resveratrol has been demonstrated to mimic metabolic activities of caloric restriction by targeting selective mechanisms of nutrient utilization and mitochondrial oxidative function to regulate energy balance in cell culture studies and animal experiments. In non-human primates, resveratrol has been shown to induce satiety signals and to reduce body-mass gain by increasing the resting metabolic rate.

One of the major metabolic target sites of resveratrol is SIRT1. By allosteric activation of SIRT1, a mammalian histone deacetylase, resveratrol enhances mitochondrial biogenesis, oxidative phosphorylation and oxygen consumption in muscle, thereby improving the mitochondrial metabolic efficiency. Apart from activation by SIRT1, resveratrol is known to elevate the catalytic activity of 5' adenosine monophosphate-activated protein kinase (AMPK), leading to reduced gluconeogenesis and lipogenesis.

This review provides an overview of the mechanistic action of resveratrol on various pathways regulating satiety and energy metabolism.

[#] This chapter was previously published in *Resveratrol: Sources, Production and Health Benefits,* edited by Dominique Delmas, New York: Nova Publishers Inc., 2013.

[*] Tel: +43-1-4277-706 10 Fax: +43-1-4277-9 706. E-mail: veronika.somoza@univie.ac.at.

In: Biochemistry Research Summaries. Volume 1
Editors: James T. Walker and Catherine M. Jones
ISBN: 978-1-53610-895-8

Chapter 27

RESVERATROL: A POTENT AGENT AGAINST INFLAMMATORY AND OXIDATIVE STRESS-RELATED DISEASES[#]

Carmen Pinto, Beatriz Rodríguez-Galdón and Pedro Macías*
Department of Biochemistry and Molecular Biology,
Faculty of Sciences, Extremadura University, Badajoz, Spain

RESEARCH SUMMARY

One of the major ingredients of the Ko-jo-kon, a Japanese folk medicine applied to several inflammations, is the root of the plant *Polygonum cuspidatum*, where resveratrol was identified for the first time. The protective role of resveratrol against inflammatory and stress oxidative- related diseases is mediated by its role as a modulator of inflammatory redox signaling pathways by interaction with the nuclear factor, NF-κB, whose activation is critical in numerous inflammatory pathologies. One of the early biochemical studiesy, dated at 1985, in which the effect of stilbenes on aan oxidative metabolic pathway was studied, deals with the effect of resveratrol on arachidonic acid metabolism, showing the potent inhibitory effect on lipoxygenase and ciclooxygenase activities. Afterwards, based on studies related with platelets aggregation and synthesis of thromboxanes, it was suggested the role of resveratrol as an inhibitor of the oxidative metabolism of arachidonic acid was suggested. In animals, the main enzymes responsible for this process are cyclooxygenase and lipoxygenase. Both enzymes are involved in oxidative pathways responsible for a lot of inflammatory processes, which are strongly inhibited by resveratrol. In platelets, resveratrol mainly inhibited cyclooxygenase, in contrast, in neutrophils, is a potent inhibitor of 5- and 15- lipoxygenases. Also, resveratrol affects the arachidonic acid mobilization and the induction of cyclooxygenase in stress oxidative conditions. The inhibitory effect of resveratrol on lipoxygenase is of great interest. In animals, this enzyme, produces the synthesis of

[#] This chapter was previously published in *Resveratrol: Sources, Production and Health Benefits,* edited by Dominique Delmas, New York: Nova Publishers Inc., 2013.

* E-mail: pedrom@unex.es.

hydroperoxiderivatives of arachidonic acid, being these compounds precursors of a series of active molecules, such as leukotrienes or lipoxins, involved in relevant pathophysiological processes including inflammation, asthma, atherosclerosis and cancer.

In: Biochemistry Research Summaries. Volume 1
Editors: James T. Walker and Catherine M. Jones
ISBN: 978-1-53610-895-8

Chapter 28

EVIDENCE TO SUPPORT TESTING OF RESVERATROL FOR EFFICACY IN RHEUMATOID ARTHRITIS [#]

Paul J. Davis[1,2,3], Ole Vang[4*], Ran Meng,[1,2*] Dana Crawford[3], Jenn-Haung Lai[5], Feng-Cheng Liu[5], Heng-Yuan Tang[1,2] and Hung-Yun Lin[1,2]

[1]Ordway Signal Transduction, New York, US
[2]Albany College of Pharmacy, Albany, New York, US
[3]Albany Medical College, Albany, New York, US
[4]Department of Science, Systems and Models, Roskilde University, Roskilde, Denmark
[5]National Defense Medical Center, Neihu, Taipei, Taiwan

RESEARCH SUMMARY

Rheumatoid arthritis (RA) is a systemic autoimmune disease characterized by chronic inflammation of multiple joints with the potential to cause joint destruction and functional disability. While the effects can be seen in many tissues and organs, attacks are primarily against the synovial joints. Resveratrol, a phytoalexin found in grape skins and other plants, has been shown to exhibit anti-proliferative, anti-inflammatory, and anti-catabolic properties. Studies have suggested that these properties can be harnessed to reduce the effects associated with rheumatoid arthritis. This review will focus on the mechanisms involved in the development of rheumatoid arthritis, and the inhibitory effects of resveratrol. The risk factors involved in rheumatoid arthritis such as elevated cytokines: tumor necrosis factor-α (TNF-α), and Interleukin 1β (IL-1β); activation of the NF-κB signaling cascade; and enhanced cyclooxygenase-2 (COX-2) activity will be addressed. Mechanisms of resveratrol-induced inhibition of nuclear factor-κB (NF-κB) activation and IL-1β production will be discussed, as

---

[#] This chapter was previously published in *Resveratrol: Sources, Production and Health Benefits,* edited by Dominique Delmas, New York: Nova Publishers Inc., 2013.

[*] These authors contributed equally to this article.

well as resveratrol's induction of COX-2 nuclear accumulation and subsequent apoptosis. Resveratrol induces nuclear accumulation of COX-2 facilitating apoptosis.

Finally, the potential role of inducible COX-2-dependent apoptosis in arthritis relief will be investigated.

In: Biochemistry Research Summaries. Volume 1
Editors: James T. Walker and Catherine M. Jones
ISBN: 978-1-53610-895-8

Chapter 29

EFFECTS OF MICRORNAS IN THE BIOLOGICAL RESPONSE OF RESVERATROL[#]

Malek Kechrid and Partha Mukhopadhyay
Laboratory of Physiologic Studies, NIAAA, National Institute of Health, Bethesda, Maryland, US

RESEARCH SUMMARY

Resveratrol is a naturally occurring phenol belonging to Stilbene family of compounds. Experimental evidence shows that resveratrol is a cardio- protective agent, beneficial for atherosclerosis, hypertension, heart failure, ischemic heart disease, hypertrophy, diabetes, and obesity. It also can induce several longetivity genes including Sirt1. Substances such as resveratrol can stimulate the production of miRNAs that are synthesized by RNA polymerase as poly- adenylated pri-miRNA. Genes for miRNA are an essential component of the genetic program of all species where most of them are evolutionarily conserved.

miRNAs target 5'UTR or coding regions of mRNA, inhibit protein synthesis, and occasionally activates translation. Dicer is involved in the maturation of miRNA, knock- out of Dicer leads to cardiomyopathy and heart failure, as well as abdominal immune response and inflammation. In cancer, there are two main strategies to target miRNA expression, direct and indirect. The direct strategy involves the use of oligonucleotides or virus-based constructs to block the target miRNA. The indirect strategy involves the use of drugs to modulate miRNA expression by targeting their transcription. The results show that there are significant regulations of the gene, and they confirm that resveratrol can protect the ischemic heart by restoring the IR- induced up- regulation or down- regulation of gene expression.

[#] This chapter was previously published in *Resveratrol: Sources, Production and Health Benefits,* edited by Dominique Delmas, New York: Nova Publishers Inc., 2013.

In: Biochemistry Research Summaries. Volume 1
Editors: James T. Walker and Catherine M. Jones
ISBN: 978-1-53610-895-8

Chapter 30

CAN WE PREVENT CANCER? THE ROLE OF RESVERATROL IN CHEMOPREVENTION[#]

Eduardo Cremonese Filippi-Chiela[1], Guido Lenz[1,2*] and Lauren Lúcia Zamin[3•]
[1]Department of Biophysics
[2]Center of Biotechnology; Universidade Federal do Rio Grande do Sul (UFRGS), Porto Alegre, RS, Brazil
[3]Universidade Federal da Fronteira Sul (UFFS), Cerro Largo, RS, Brazil

RESEARCH SUMMARY

Cancer is the second cause of cell death in developed countries, losing only to cardiovascular diseases. Despite improvements in several fronts, the quality of life and overall survival to several cancer types remains largely unaltered. Thus, for several cancer types, prevention, mainly by reducing exposure to cancer inducing agents such as smoking or HPV, is the best way to reduce prevalence and incidence. However, compounds that act as chemopreventive agents have started to be noted in cancer research, and Resveratrol (Rsv) is one of the most promising molecules with potential anti-cancer effects. Chemoprevention exerted by Rsv occurs through reduction of events that have been increasingly described as pro-carcinogenic, mainly inflammation and oxidative stress. The effect of Rsv in this mechanisms occurs directly (*i.e.* acting as an antioxidant molecule or inhibiting inflammation in several steps) and indirectly (for instance, acting as an anti-obesity compound reducing, thus, harmful consequences of fat accumulation such as increase of reactive species and pro-inflammatory markers). Moreover, Rsv is able to reactivate, in cancer cells, some endogenous anticancer mechanisms, among which are apoptosis and senescence. This reactivation may kill the existing cancer cells (characterizing Rsv as an anti-progression compound) but, more importantly, help to eliminate pre-cancer cells, characterizing it as an anti-initiation

---

[#] This chapter was previously published in *Resveratrol: Sources, Production and Health Benefits,* edited by Dominique Delmas, New York: Nova Publishers Inc., 2013.

* Tel: +55 51 3308 7620; Fax +55 51 3308 7309 E-mail: lenz@ufrgs.br.

compound. In this chapter, we will discuss the *in vitro* and *in vivo* models of chemopreventive action of Rsv, focusing in its inferred mechanism of action.

In: Biochemistry Research Summaries. Volume 1
Editors: James T. Walker and Catherine M. Jones

ISBN: 978-1-53610-895-8

Chapter 31

THERAPEUTIC POTENTIAL OF RESVERATROL AGAINST CANCERS: FOCUS ON CELL PROLIFERATION AND CELL DEATH PATHWAYS[#]

Dominique Delmas[1,2], Virginie Aires[1,2], Emeric Limagne[1,3], Gérard Lizard[1,3], Aziz Hichami[1,2], François Ghiringhelli[1,2] and Norbert Latruffe[1,3]

[1]Université de Bourgogne, Dijon, France
[2]Inserm Research Center, Team Chemotherapy, Lipid Metabolism and Antitumoral Immune Response, Dijon, France
[3]Inserm Research Center, Team Metabolic and Nutritionnal Biochemistry, Dijon, France

RESEARCH SUMMARY

Cancers are the largest cause of mortality and morbidity in industrialized countries. Several new concepts have emerged in relation to mechanisms that contribute to the regulation of carcinogenesis processes and associated inflammatory effects. One of the most potent defences against cancer is the induction of programmed cell death, also called apoptosis, in tumor cells leading in eliminating potentially deleterious mutated cells. Hence, the main target for anticancer therapeutic strategies against malignant diseases is the reestablishment of apoptotic homeostasis, induction of tumor cell apoptosis and overcoming resistance to current cytotoxic therapies. In the field of the research of new compounds able to act on these pathways, dietary polyphenols (such as resveratrol and derivatives) appear to be of great interest. Indeed, numerous studies have reported interesting properties of *trans*-resveratrol, a natural trihydroxystilbene found in various food products, as a chemopreventive agent against important pathologies such as cardiovascular, neurodegenerative and immune diseases, and cancers. The main molecular mechanism of resveratrol action appears to be its

[#] This chapter was previously published in *Resveratrol: Sources, Production and Health Benefits,* edited by Dominique Delmas, New York: Nova Publishers Inc., 2013.

ability to trigger cell cycle arrest, to promote cell death by apoptosis, and to modulate some miRNAs expression. This review focuses on the pathways mediating resveratrol-induced tumor cell proliferation arrest. Resveratrol disturbs the cell cycle progression through a modulation of the key regulators such as cyclins and their partners cyclin dependent-kinases. We show and discuss the relationship between the resveratrol-induced tumor cell accumulation during DNA replication phase and the chemosensitization to various cytotoxic drugs. Moreover, resveratrol is able to induce cell death via both mitochondria and death receptors pathways. In this review, we also analyze the intrinsic and extrinsic death pathways promoted by this polyphenol, and we present how resveratrol can act as a chemosensitizer agent by lowering cell death threshold induced by classical anticancer drugs and by counteracting tumor cell chemoresistance.

In: Biochemistry Research Summaries. Volume 1
Editors: James T. Walker and Catherine M. Jones
ISBN: 978-1-53610-895-8

Chapter 32

ROLE OF DNA DAMAGE IN THE ANTITUMORAL ACTIVITY OF RESVERATROL[#]

Emiliano Basso, Stefano Leone and Renata Cozzi*
Department of Biology, Roma Tre University, Viale Marconi, Roma, Italy

RESEARCH SUMMARY

Resveratrol (RESV) is a natural molecule detected in at least 70 plant species, a number of which are normal component of human diet. The interest in RESV is growing in a exponential manner since its chemopreventive activity in all three stages of carcinogenesis was demonstrated. Afterward a great number of studies indicated that RESV is implicated in various cellular processes and exerts several beneficial effects on human health. However, controversial data are present in literature about the ability of RESV to affect DNA molecule integrity both in cancer and normal cells. RESV seems to induce DNA breaks through both a pro-oxidant action and an inhibition of topoisomerase enzymes. The consequence of these activities is the induction of a delay in cell cycle progression which can evolve in senescence and/or cell death. On the other hand recent reports are dealing with the ability of RESV to protect towards DNA damage induced by exogenous and endogenous exposure. At moment no definitive conclusion has been reached on this subject, but rather it states the need to collect more data to demonstrate the efficacy of RESV against tumor cells coupled with a virtual absence of toxicity in non-proliferating normal cells.

[#] This chapter was previously published in *Resveratrol: Sources, Production and Health Benefits,* edited by Dominique Delmas, New York: Nova Publishers Inc., 2013.

[*]E-mail: cozzi@uniroma3.it.

In: Biochemistry Research Summaries. Volume 1
Editors: James T. Walker and Catherine M. Jones
ISBN: 978-1-53610-895-8

Chapter 33

RESVERATROL AND NEUROPROTECTION#

Carmem Gottfried[1*], André Quincozes-Santos[1], Kader Basli[2] and Tristan Richard[2]

[1] Department of Biochemistry, Postgraduate Program of Biochemistry, Institute of Basic Health Sciences, Federal University of Rio Grande do Sul, Porto Alegre, RS, Brazil
[2] Univ. de Bordeaux, ISVV, GESVAB, Villenave d'Ornon, France

RESEARCH SUMMARY

One of the challenges in medicine research is the development of safe drugs for the treatment of brain pathologies, attempting to prevent neuronal degeneration, and loss of function. As redox imbalance is implicated in the pathogenesis of a number of neurological disorders such as Alzheimer's disease, Parkinson's disease, multiple sclerosis and stroke, there is a growing interest in the antioxidant resveratrol (RSV), a polyphenol derived from plants. Beyond the RSV ability to modulate the expression of genes and proteins, recently evidence suggests that it is also able to modulate cell signaling within cells, targeting different proteins and enzymes. The neural regulation by RSV at different levels could be an explanation of some of its neuroprotective actions, as well as its influence on blood flow, cell death, and inflammatory cascades. Both experimental and clinical results have shown that RSV is a promising neuroprotectant in reducing brain damage and tumor in complex manner, involving neuroglial plasticity and blood-brain barrier modulation. The present chapter summarizes the *in vitro* and *in vivo* experimental results highlighting the possible role of RSV as neuroprotective biofactor, with particular attention to its neurovascular, antioxidant, antiinflamatory, anti-neurodegenerative and antitumoral effects in brain.

This chapter was previously published in *Resveratrol: Sources, Production and Health Benefits,* edited by Dominique Delmas, New York: Nova Publishers Inc., 2013.

* E-mail: cgottfried@ufrgs.br.

In: Biochemistry Research Summaries. Volume 1
Editors: James T. Walker and Catherine M. Jones

ISBN: 978-1-53610-895-8

Chapter 34

RESVERATROL AND NEURODEGENERATIVE DISEASES[#]

Altaf S. Darvesh[1,2*] and Anupam Bishayee[3]
[1]Department of Pharmaceutical Sciences, College of Pharmacy,
Northeastern Ohio Medical University, Rootstown, Ohio, US
[2]Department of Psychiatry, College of Medicine,
Northeastern Ohio Medical University, Rootstown, Ohio, US
[3]Department of Pharmaceutical and Administrative Sciences, School of Pharmacy,
American University of Health Sciences, Signal Hill, California, US

RESEARCH SUMMARY

Resveratrol (3,4´,5-trihydroxy-*trans*-stilbene), a polyphenolic compound of dietary origin, is found in several plant sources such as berries, grapes, and peanuts. In the past two decades, resveratrol has been a subject of intense investigation for its therapeutic potential in a plethora of age-related chronic ailments, such as cardiovascular, neoplastic and neurodegenerative diseases. Biochemical studies reveal that the strong antioxidant and anti-inflammatory property of resveratrol as well as its fascinating ability to modulate a wide array of signaling pathways implicated in the pathophysiology of the aforementioned illnesses. Both oxidative stress and inflammation contribute to the neuronal loss and neurodegeneration observed in serious illnesses, such as Alzheimer's disease and Parkinson's disease along with other deleterious factors, such as apoptotic and necrotic damage and protein aggregation. In this chapter, we review the current understanding regarding the potential use of resveratrol in the prevention and treatment of Alzheimer's disease and Parkinson's disease. Pre-clinical investigations on the pharmacological effects of resveratrol in both *in vitro* and *in vivo* models of Alzheimer's disease and Parkinson's disease are systematically reviewed. Although deemed promising, the clinical application of resveratrol is beset with several roadblocks,

---

[#] This chapter was previously published in *Resveratrol: Sources, Production and Health Benefits,* edited by Dominique Delmas, New York: Nova Publishers Inc., 2013.

[*] Tel: 330-325-6658, Fax: 330-325-5936, Email: adarvesh@neomed.edu.

such as poor bioavailability, lack of sufficient clinical evidence and dosing studies. We review the current progress in areas, such as development of semi-synthetic stilbenoids as well as novel formulations of resveratrol and its analogs. Finally, novel hypothesis, innovative concepts and future directions required to successfully transition reseveratrol from an alternative treatment option to a robust mainstream neuroprotective agent with a well-defined therapeutic profile in the treatment of Alzheimer's disease, Parkinson's disease as well as other neurodegenerative disorders is critically discussed.

In: Biochemistry Research Summaries. Volume 1
Editors: James T. Walker and Catherine M. Jones

ISBN: 978-1-53610-895-8

Chapter 35

SOURCES AND FOOD MATRIX ON THE BIOAVAILABILITY OF RESVERATROL[#]

Maria Rotches-Ribalta[1,2*], Mireia Urpi-Sarda[3,4], Sara Tulipani[1,2], Pilar Trespalacios[1,2], Olha Khymenets[1,2], Montse Rabassa-Bonet[1,2], Maria Boto-Ordóñez[3,4], Rosa Vazquez-Fresno[1,2], Mar Garcia-Aloy[1,2], Rafael Llorach[1,2] and Cristina Andres-Lacueva[1,2†]

[1] University of Barcelona, Av Joan XXIII, s/n, Barcelona, Spain
[2] Ministry of Science and Innovation, Barcelona, Spain
[3] Institut d'Investigació Biomèdica August Pi i Sunyer, Barcelona, Spain
[4]Instituto de Salud Carlos, Barcelona, Spain

RESEARCH SUMMARY

Resveratrol consumption has been positively associated with the prevention of a number of diseases, mainly due to its reported antioxidant activity and anti-inflammatory and antitumor effects, and has, therefore, the potential to be evaluated for therapeutic use. However, scarce studies have been carried out in humans and the results yielded are still controversial. Before making nutritional recommendations regarding this promising molecule, other aspects such as dietary intake and bioavailability should be considered.

[#] This chapter was previously published in *Resveratrol: Sources, Production and Health Benefits,* edited by Dominique Delmas, New York: Nova Publishers Inc., 2013.

[*1] Nutrition and Food Science Department, XaRTA-INSA. Pharmacy Faculty, University of Barcelona, Av Joan XXIII, s/n, Barcelona, Spain.
[2] Ingenio-CONSOLIDER program, FUN-C-FOOD, Spain CSD2007-063. Ministry of Science and Innovation, Barcelona, Spain.
[3] Department of Internal Medicine, Hospital Clinic, Institut d'Investigació Biomèdica August Pi i Sunyer (I. D. I. B. A. P. S.), Villarroel 170, 08036 Barcelona, Spain.
[4] CIBER 06/03: Fisiopatología de la Obesidad y la Nutrición, Instituto de Salud Carlos III.
[†] E-mail: *candres@ub.edu.*

Resveratrol and other stilbenes, such as piceid, have been identified in a large number of plants, although they have been found in a few dietary components, with red wine and grape products as the major sources. Therefore, urinary resveratrol metabolites have been described as nutritional biomarkers of wine consumption, providing an additional tool for clinical and epidemiological studies, since they reflect a more objective assessment of nutrient intake than self-reported dietary data. Its estimated intake or exposure is essential in order to relate resveratrol consumption with its possible health effects *in vivo*.

However, resveratrol is rapidly metabolized, and therefore the bioactivity of this compound will be related to its metabolites. Identification and measurement of metabolic forms are key prerequisites for understanding the role of these compounds, since they will be the ones that reach the peripheral circulation and tissues where they could exert their biological effect. Phase II metabolism of resveratrol has been widely studied, with the glucuronide and sulfate forms being the main resveratrol conjugates.

However, few studies have focused on the microbial metabolism of resveratrol, which leads to dihydroresveratrol formation, although this could give a better idea about effects of long-term resveratrol consumption. Furthermore, the matrix in which resveratrol is delivered could play an important role in its bioavailability.

This chapter, therefore, provides an overview of the metabolism of resveratrol after short- and long-term consumption and the microbiota and food matrix effects on its bioavailability. This needs to be considered in order to gain a better understanding of the links between resveratrol intake and health.

In: Biochemistry Research Summaries. Volume 1
Editors: James T. Walker and Catherine M. Jones
ISBN: 978-1-53610-895-8

Chapter 36

PHARMACOLOGICAL ACTIVITIES, PHARMACOKINETICS AND DELIVERY APPROACHES#

Surajit Das[1,*] and Anumita Chaudhury[2,#]
[1]Institute of Chemical and Engineering Sciences, Singapore, Republic of Singapore
[2]School of Applied Science, Singapore, Republic of Singapore

RESEARCH SUMMARY

Resveratrol (3, 5, 4'-trihydroxystilbene) is a naturally occurring phytoalexin present in a wide variety of plants, such as grapes, peanuts, and mulberries. It displays wide pharmacological activities and is known for its antioxidant, antiinflammatory, analgesic, cardioprotective, hepatoprotective, neuroprotective, chemopreventive, chemotherapeutic, and antiaging activities. In addition, resveratrol has shown promising therapeutic efficacy towards several lower gastro-intestinal (GI) tract diseases like, colorectal cancer and colitis. However, most of the success was reported in *in vitro* disease models. As the pharmacology of resveratrol was subjected to extensive studies during the past decade, its pharmacokinetics has also been investigated in pre-clinical models as well as in humans. Unfortunately, the pharmacokinetic properties of resveratrol are not as favourable as its beneficial pharmacological activities in various disease models. In spite of rapid absorption through GI tract, resveratrol has exhibited very short plasma half-life. Upon administration, resveratrol undergoes extensive phase II metabolism to from glucuronide and sulfate conjugates. Most importantly, its oral bioavailability in animal and human was found to be very low. The poor oral bioavailability and extremely short plasma half-life of resveratrol have raised concerns regarding its' systemic action. Hence, improved delivery approaches of resveratrol are required to transform its huge *in vitro* success to *in vivo* benefits. In search of the improved

---

This chapter was previously published in *Resveratrol: Sources, Production and Health Benefits*, edited by Dominique Delmas, New York: Nova Publishers Inc., 2013.

* Institute of Chemical and Engineering Sciences, A*STAR (Agency for Science, Technology and Research), 1 Pesek Road, Jurong Island, Singapore 627833, Republic of Singapore. E-mail: *surajit_das@ices.a-star.edu.sg; surajitdas1982@yahoo.com.*

School of Applied Science, Republic Polytechnic, 9 Woodlands Ave 9, Singapore 738964, Republic of Singapore.

local/systemic action, researchers have developed different drug delivery techniques. Resveratrol was solubilized by cyclodextrins and administered through oral, intravenous and topical routes. Other formulations, such as topical gel, nanosponge, liposphere, emulsion, liposome, microcapsule, micelle, nanoparticle, etc. also have been exploited to improve local/systemic delivery of resveratrol. Co-administration with other compounds, such as curcumin, quercetin has also been tried to improve resvertrol's efficacy by reducing metabolism or improving uptake. Another delivery approach has been attempted for the treatment of the colon diseases (colonic inflammation and colorectal cancer). Because of the extensive metabolism and rapid absorption through the upper GI tract, insufficient amount of resveratrol reaches to the lower GI tract. Thus, different delayed release and colon-specific delivery systems have been investigated to provide a potential approach to treat colonic diseases. This kind of delivery systems protect drug during its transit through the upper GI tract and allow its release in the colonic region, which should be able to bypass the rapid absorption and metabolism issues of resveratrol. Different microparticles/beads based on biopolymer pectin were developed for the delayed release and colon-specific delivery of resveratrol, which demonstrated *in vitro* and *in vivo* colon-specific release of resveratrol.

In: Biochemistry Research Summaries. Volume 1
Editors: James T. Walker and Catherine M. Jones

ISBN: 978-1-53610-895-8

Chapter 37

RESVERATROL AND CYCLODEXTRINS: RECENT ADVANCES IN ENCAPSULATION#

C. Lucas-Abellán, M. I. Fortea, M. T. Mercader-Ros, A. Martínez-Cachá, J. A. Gabaldón and E. Núñez-Delicado*
Universidad Católica San Antonio de Murcia, Murcia, Spain

RESEARCH SUMMARY

Resveratrol (3,5,4´-trihydroxy-*trans*-stilbene), which naturally occurs in peanuts, grapes, and grape-derived foodstuffs such as red wine, has been reported to exert many different health-promoting effects including antioxidant, anti-inflamatory, antitumor, antiplatelet aggregation, cardioprotective and aging-delay effects. However, despiste the huge number of studies regarding the health-beneficial effects of resveratrol, its effectiveness from a dietary point of view is not clear.

The dietary intake can be increased through resveratrol-enriched functional foods. The use of resveratrol as functional ingredient in novel foods (as a fortifier and nutraceutical compound) has been restricted to date. The high hydrophobicity of resveratrol and its sensitivity to external agents such us air, light and oxidative enzymes may constitute a serious problem for its bioavailability, formulation and manipulation in the elaboration of functional foods. In order to avoid this limitation, inclusion of resveratrol in cyclodextrins has recently been suggested by our group. Cyclodextrins (CDs) are a group of naturally occurring cyclic oligosaccharides derived from starch with six, seven or eight glucose residues linked by α(1-4) glycosidic bonds in a cylinder-shaped structure, and denominated α-, β- and γ-CDs, respectively. The central cavity of these molecules is hydrophobic, while the rims of the surrounding walls are hydrophilic. This hydrophobic cavity forms inclusion complexes with resveratrol, increasing its solubility, stability, bioavailability and antioxidant activity. The

---

This chapter was previously published in *Resveratrol: Sources, Production and Health Benefits,* edited by Dominique Delmas, New York: Nova Publishers Inc., 2013.

* Dpto. de Tecnología de la Alimentación y Nutrición. Universidad Católica San Antonio de Murcia. Avenida de los Jerónimos s/n. 30107 Guadalupe. Murcia. España. E-mail: *enunez@pdi.ucam.edu.*

complexation of resveratrol in CDs increase its antioxidant activity, with CDs acting as a controlled dosage reservoir that protects resveratrol against its rapid oxidation by free radicals.

Future research should focus on translating in vitro findings regarding potential benefits into in vivo models. In addition, further understanding of the potential toxicity, health effects, bioavailability, and metabolism of resveratrol is necessary before dietary and supplements recommendations can be made. Scientific information summarized in this review supports the many potential health benefits of resveratrol and the future mechanisms for the functional foods and nutraceuticals elaboration.

In: Biochemistry Research Summaries. Volume 1
Editors: James T. Walker and Catherine M. Jones

ISBN: 978-1-53610-895-8

Chapter 38

CLINICAL DEVELOPMENT OF RESVERATROL[#]

Ketan R. Patel, William P. Steward, Andreas J. Gescher and Karen Brown*
Department of Cancer Studies and Molecular Medicine, University of Leicester, Leicester, UK

RESEARCH SUMMARY

Mounting preclinical data support the concept that consumption of resveratrol can offer numerous health benefits for humans. Evidence suggests that it may play a role in the management of a wide variety of diseases including cardiovascular, diabetes and cancer.Encouragingly, doses of up to 5 g/day, taken for a month, have been show to be safe and reasonably well tolerated. However, the occurrence of dose-related mild to moderate gastrointestinal side effects and the ability of resveratrol to interfere with drug metabolizing enzymes are likely to dictate the upper dosage limit and it has been recommended that future studies be restricted to doses of less than 1.0 g/day. Further assessment of the long-term toxicity of resveratrol is needed at doses within this range. In humans, as in rodent models, resveratrol is efficiently absorbed when taken orally but rapid phase II metabolism drastically limits its plasma bioavailability and this may also be the case for tissues that are distant to the gastrointestinal tract. It is important to exploit the clinical pharmacokinetic data now available to guide the range of concentrations used in preclinical *in vitro* experiments. Currently there are no published clinical reports indicating therapeutic or protective effects of resveratrol and only a handful of studies have attempted to gain evidence of pharmacological activity in terms of pharmacodynamic changes in humans. However, the research focus is now clearly shifting towards the assessment of efficacy and the mechanisms engaged in humans and results emanating from the many trials that are either scheduled or ongoing should begin to address this knowledge gap. This chapter summarises the available data from published

[#] This chapter was previously published in *Resveratrol: Sources, Production and Health Benefits,* edited by Dominique Delmas, New York: Nova Publishers Inc., 2013.

* E-mail: kb20@le.ac.uk.

human studies and describes how it supports the further development of resveratrol for disease prevention and treatment.

In: Biochemistry Research Summaries. Volume 1
Editors: James T. Walker and Catherine M. Jones
ISBN: 978-1-53610-895-8

Chapter 39

RESVERATROL DERIVATIVES AND HEALTH BENEFITS[#]

Simone Fulda*
Institute for Experimental Cancer Research in Pediatrics,
Goethe-University Frankfurt, Frankfurt, Germany

RESEARCH SUMMARY

Resveratrol is a naturally occurring phytoalexin that exerts a multitude of beneficial effects on human health and is widely present in foods and beverages. As far as cancer is concerned, resveratrol has been implicated in both chemoprevention and cancer therapy. A large variety of resveratrol analogs exists, which either naturally occur or which have been synthesized. Structure-activity studies are instrumental to optimize such compounds for their application in human diseases. Therefore, the future development of resveratrol derivatives is expected to provide new pharmacological tools for chemoprevention and treatment of cancer.

[#] This chapter was previously published in *Resveratrol: Sources, Production and Health Benefits,* edited by Dominique Delmas, New York: Nova Publishers Inc., 2013.
[*] Tel.: +49 69 67866557; Fax: +49 69 6786659157; E-mail: simone.fulda@kgu.de.

In: Biochemistry Research Summaries. Volume 1
Editors: James T. Walker and Catherine M. Jones
ISBN: 978-1-53610-895-8

Chapter 40

MOLECULAR COMPLEXATION OF GLYCINE BY 18-CROWN-6 IN AQUEOUS – ORGANIC SOLVENTS: A SOLVATION -THERMODYNAMIC STUDY[#]

T. R. Usacheva[1*], V. A. Sharnin[1] and E. Matteoli[2,±]
[1]Ivanovo State University of Chemistry and Technology, Ivanovo, Research Institute of Thermodynamics and Kinetics of Chemical Processes, Russia
[2]CNR-IPCF, Institute of Physical-Chemical Processes of National Research Council, Pisa, Italy

RESEARCH SUMMARY

In this chapter we discuss the influence of water-ethanol (H_2O-EtOH), water-dimethylsulfoxide (H_2O-DMSO) and water–acetone (H_2O-MeAc) mixed solvents on the molecular complex formation reaction between biological model substances glycine (Gly) and 18-crown-6 (18C6), 18C6+Gly ↔ [Gly18C6]. The analysis of the interdependence between the thermodynamic reaction parameters (Gibbs energy, enthalpy and entropy) of complex formation and the thermodynamic parameters for the solvation of reagents allows to reveal regularities and peculiarities of the influence of solvent composition on formation of [Gly18C6].

An increase of lgK^o([Gly18C6]) and of exothermicity for $\Delta_r H^\circ$([Gly18C6]) with increasing the non-aqueous component mole fraction in mixed solvents is observed as a common effect for all studied media. The discussion and interpretation of this phenomenon is based on the solvation-thermodynamic approach.

The analysis of the solvation enthalpy contributions of all reaction participants, $\Delta_{tr}H^\circ$(Gly), $\Delta_{tr}H^\circ$(18C6), and $\Delta_{tr}H^\circ$([Gly18C6]), to the reaction enthalpy transfer, $\Delta_{tr}H^\circ_r$, shows that the strengthening of the desolvation of 18C6 and Gly with increasing the

[#] This chapter was previously published in *Glycine: Biosynthesis, Physiological Functions and Commercial Uses*, edited by Wilhelm Vojak, New York: Nova Publishers Inc., 2013.

[*] Email:oxt@isuct.ru.
[±] Email:matteoli@ipcf.cnr.it.

concentration of EtOH, DMSO or MeAc in the mixed solvents favors the increase of the reaction exothermicity. The thermodynamic ratio α_{dif} determined by means of the solvation-thermodynamic approach is used for estimating the enthalpy changes for the complexation reaction in the studied solvents according to: $\Delta_{tr}H^{\circ}_{r,\ calc} = (\alpha_{dif} - 1)\Delta_{tr}H^{\circ}(18C6)$. The good agreement between $\Delta_{tr}H^{\circ}_{r.calc}$ and $\Delta_{tr}H^{\circ}_{r}$ in the examined composition range of the mixed solvents allows to estimate the influence of H_2O-EtOH, H_2O-DMSO, H_2O-MeAc solvent composition on the energetics of complex formation at the solvent compositions with high concentration of non-aqueous components where the very low solubility of glycine makes impossible the experimental determination of the reaction enthalpy. The relationship between the solvation contributions of the reagents for the transfer of reaction Gibbs energy from water into H_2O-EtOH and H_2O-DMSO solvents is essentially different from the corresponding values of the reaction enthalpy. The analysis of the solvation contributions of reagents into the $\Delta_{tr}G^{o}_{r}$ in H_2O-EtOH and H_2O-DMSO solvents reveals the prevailing role of the desolvation contribution of Gly, $\Delta_{tr}G^{o}$(Gly), in determining the Gibbs energy of transfer of reaction from water in water-organic solvents. The behavior of the thermodynamic parameters of molecular complex formation reaction with 18C6 of glycine is compared with that of D,L-alanine and L-phenylalanine as well as with thermodynamic data of complexation between 18C6 and silver (I) ion in non aqueous media.

In: Biochemistry Research Summaries. Volume 1
Editors: James T. Walker and Catherine M. Jones
ISBN: 978-1-53610-895-8

Chapter 41

ONE-POT ELECTROSYNTHESIS OF POLYPEPTIDES ON SURFACES IN CONCENTRATED AMINO ACID BASED ELECTROLYTES: STUDY OF CONCENTRATED GLYCINE BASED ELECTROLYTES AND APPLICATION TO SOLID pH SENSOR[#]

Guillaume Herlem[1,*], Hatem Boulhadour[1], Alexandros Antoniou[2], Freddy Torrealba Anzola[3] and Tijani Gharbi[1]

[1]Nanosciences Lab, Imagery and Therapeutics, UFR ST, CHU Jean Minjoz, University of Franche-Comte, France
[2]Technological Institute of Larissa, Department of Medical Laboratories, Greece
[3]UICM-Departamento de Física, UCLA-Decanato de Ciencias y Tecnología, Venezuela

RESEARCH SUMMARY

The anodic oxidation of concentrated based glycine electrolyte on smooth platinum electrode yields a strongly grafted polyglycine-like coating. This electrochemical reaction occurs in an irreversible way on platinum but also on other material such as graphite. Contrary to studies performed in diluted medium, spectroscopic surface analysis revealed the presence of peptide bonds in the resulting thin film layer on platinum with polyglycine like (type II) structure. Since the electrodeposited polyglycine like coating at alkaline pH have chemical groups with affinity towards protons, it was used as transducer and embedded in a solid potentiometric pH sensor. Local pH measurement can be achieved by using

---

[#] This chapter was previously published in *Glycine: Biosynthesis, Physiological Functions and Commercial Uses*, edited by Wilhelm Vojak, New York: Nova Publishers Inc., 2013.

[*] To whom all correspondence should be addressed: Email: guillaume.herlem@univ-fcomte.fr.

miniaturized electrodes thanks to photolithography process. The couple (silver chloride as reference electrode, polyglycine based platinum electrode as working electrode) of microelectrodes gives linear potentiometric responses versus pH in the range [2 - 12], reversibly and with a sensitivity of 52.4 mV/pH (for 1 mm electrode diameter). This solid pH sensor was tested over a period of thirty days and its potentiometric behavior was compared to other polymer based pH electrodes such as linear polyethylenimine (L-PEI) and polyaniline (PANI). Quality control tests can be considered by combining miniaturized pH sensors with other measurements.

In: Biochemistry Research Summaries. Volume 1
Editors: James T. Walker and Catherine M. Jones
ISBN: 978-1-53610-895-8

Chapter 42

HETEROLOGOUS EXPRESSION OF THE GLYCINE RECEPTOR CHLORIDE CHANNEL[#]

Qiang Shan[1,*] *and Lu Han*[2,#]
[1]Brain and Mind Research Institute, University of Sydney, Sydney, NSW, Australia
[2]Queensland Brain Institute, University of Queensland, Brisbane, QLD, Australia

RESEARCH SUMMARY

The glycine receptor (GlyR) is a neuronal chloride channel, which, upon binding glycine, allows chloride anions to permeate into the neuron causing hyperpolarization of the transmembrane potential and inhibition of neuronal activity.

To investigate the structure and function of the GlyR, scientists usually isolate it by expressing it in a heterologous expression system, such as the *Xenopus laevis* oocyte or a cultured immortalized mammalian cell line, such as HEK 293.

In this chapter, protocols for the expression of the GlyR in these two expression systems are given, including preparation of recombinant cDNA, preparation of cRNA, culturing HEK 293 cells, surgical isolation of *Xenopus laevis* oocytes, transfection of cDNA into the HEK 293 cells, injection of cRNA into the oocytes, and detection of the expressed GlyRs.

[#] This chapter was previously published in *Glycine: Biosynthesis, Physiological Functions and Commercial Uses*, edited by Wilhelm Vojak, New York: Nova Publishers Inc., 2013.

[*] Corresponding author: Qiang Shan. Brain and Mind Research Institute, University of Sydney, Sydney, NSW, 2050, Australia. E-mail: qshan@yahoo.com, Telephone: +61 2 9114 4032; Fax: +61 2 9114 4035.
[#] Lu Han: Queensland Brain Institute, University of Queensland, Brisbane, QLD, 4072, Australia.

In: Biochemistry Research Summaries. Volume 1
Editors: James T. Walker and Catherine M. Jones
ISBN: 978-1-53610-895-8

Chapter 43

GLYCINE: THEORY OF THE INTERACTION WITH FAST ION RADIATION[#]

John R. Sabin[1,2], Jens Oddershede[1,2] and Stephan P. A. Sauer[3]

[1]Departments of Physics and Chemistry, University of Florida, Gainesville, Florida, US
[2]Institute for Physics, Chemistry, and Pharmacology, University of Southern Denmark, Denmark
[3]Department of Chemistry, University of Copenhagen, Copenhagen, Denmark

RESEARCH SUMMARY

With the advent of the use of precise ion accelerators for medical purposes, it becomes ever more important to understand the interaction of biomolecules with fast ions. Glycine is both a protein component and a model biomolecule, and is thus an important test system. In this report, we discuss various aspects of the theoretical evaluation of the properties of energy deposition associated with the collision of fast ions with glycine and its zwitterions, including differences due to molecular conformation and orientation with respect to the ion beam direction as well as due to the effect of surrounding water molecules and the state of target aggregation. Quantum mechanical calculations, which yield the dipole oscillator strength distribution of glycine are reported. The ease with which energy is absorbed from a fast ion, described by the mean excitation energy and stopping power of glycine, is reported and discussed.

[#] This chapter was previously published in *Glycine: Biosynthesis, Physiological Functions and Commercial Uses*, edited by Wilhelm Vojak, New York: Nova Publishers Inc., 2013.

In: Biochemistry Research Summaries. Volume 1
Editors: James T. Walker and Catherine M. Jones
ISBN: 978-1-53610-895-8

Chapter 44

ELECTROCHEMISTRY OF ALIPHATIC BIFUNCTIONAL MOLECULES: FROM POLYMERIC THIN FILM COATING TO SENSORS AND BIOSENSOR APPLICATIONS[#]

Guillaume Herlem, Hatem Boulhadour, Alexandros Antoniou, Freddy Torrealba Anzola and Tijani Gharbi
University of Franche-Comte and CHU Jean Minjoz, France

RESEARCH SUMMARY

Despite their toxicity and the rise of green chemistry, nonaqueous solvents play an important role in analytical electrochemistry. When used methodically, they can bring an elegant way to modify a substrate irreversibly in one-pot experiment. Electrochemistry in liquid ammonia and liquid amines contributes in this direction too. Since ammoniates and aminates (by analogy to the name hydrate for aqueous electrolytes) are very used in our group, I am doing electrochemistry in these media:

1 – to elucidate new (electrochemical) reactions, electropolymerization, etc.
2 – to better discriminate the analytes by use of several techniques in (bio) chemical sensors (bio) analytical
3 - to model and interpret the results by theoretical ab initio and docking.

I will explain, in what follows, the three points that summarize my research I initiated and developed in Besancon, France.

[#] This chapter was previously published in *Glycine: Biosynthesis, Physiological Functions and Commercial Uses*, edited by Wilhelm Vojak, New York: Nova Publishers Inc., 2013.

In: Biochemistry Research Summaries. Volume 1
Editors: James T. Walker and Catherine M. Jones
ISBN: 978-1-53610-895-8

Chapter 45

CLINICAL USES OF GLYCINE[#]

Michael D. Davis[1], Brian K. Walsh[2] and John F. Hunt[3]
[1]Virginia Commonwealth University School of Nursing – Adult Health and Nursing System, Richmond, VA, US
[2]Children's Medical Center, Dallas, TX, US
[3]University of Virginia School of Medicine - Department of Pediatrics Charlottesville, VA, US

RESEARCH SUMMARY

Glycine is a non-essential amino acid used commonly as a constituent of amino-acid based nutritional supplements (enteral formulas) in medical settings. As an example, standard infant nutritional formulas all contain glycine (1-2 g/100 g protein), in an effort to match breast milk - which contains 2.7 g glycine per 100 g protein (http://www.bright beginnings.com/ -- Feb 2007), which amounts to approximately 35 to 40 mg per 5 fl oz feeding. Glycine is also used to lavage the genitourinary tract during gynecologic and urologic surgery. More recently, in the treatment of pulmonary hypertension, glycine has been used as an alkalinizing agent for inhaled prostacyclin, which requires alkaline conditions for stability. Presently, studies are evaluating the use of inhaled isotonic alkaline glycine buffer as a diluent for many inhaled therapeutic compounds and the use of intravenous glycine for amelioration of oxidative stress and ischemia/reperfusion injury.

[#] This chapter was previously published in *Glycine: Biosynthesis, Physiological Functions and Commercial Uses*, edited by Wilhelm Vojak, New York: Nova Publishers Inc., 2013.

In: Biochemistry Research Summaries. Volume 1
Editors: James T. Walker and Catherine M. Jones
ISBN: 978-1-53610-895-8

Chapter 46

DIHYDROPYRIMIDINONE DERIVATIVES: REDOX REACTIVITY, PHARMACOLOGICAL RELEVANCE AND MEDICINAL APPLICATIONS[#]

Marta Pineiro, Bruno F. O. Nascimento and António M. d'A. Rocha Gonsalves
Department of Chemistry, University of Coimbra,
Coimbra, Portugal

RESEARCH SUMMARY

Biginelli and Biginelli-like dihydropyrimidines constitute a ubiquitously recognized class of nitrogen-containing compounds. The present chapter intends to assess the currently available scientific literature on the developments specifically regarding the important oxidation/reduction reactivity of dihydropyrimidinones, their thione analogs and other derivatives. Accounts dealing with the pharmacological properties and uses in medicine of dihydropyrimidine scaffolds will also be reviewed.

[#] This chapter was previously published in *Quinones: Occurrence, Medicinal Uses and Physiological Importance*, edited by Ervin R. Price and Smith C. Johnson, New York: Nova Publishers Inc., 2013.

In: Biochemistry Research Summaries. Volume 1
Editors: James T. Walker and Catherine M. Jones
ISBN: 978-1-53610-895-8

Chapter 47

BIOLOGICAL IMPLICATIONS OF BENZOQUINONES#

Jisook Kim*
Department of Chemistry, University of Tennessee at Chattanooga, Chattanooga, TN, US

RESEARCH SUMMARY

Benzoquinones (BQs) represent the simplest form of quinones, containing two carbonyl groups on a six-membered ring. They are ubiquitously found in diverse organisms as free quinones, protein cofactors, or an integral part of the mitochondrial electron transport chain (ETC). In addition, many BQs are identified as environmental toxins generated from industrial processes as the metabolites of polycyclic aromatic hydrocarbons, contributing to bioaccumulation. To date, animal and epidemiological studies revealed that the quinone derivatives of benzene metabolites serves as a source of inducing abnormal cell behavior, leading to cancer or triggering immune response. Whether occurring endogenously in living organisms or exogenously in the environment, there is a universal understanding on the role of BQs as potential toxins, except some limited cases like protein-bound BQs or electron carriers in the ETC. Studies done at a molecular-level approach revealed that BQs exhibit both genotoxicity and non-genotoxicity/epigenetic toxicity, targeting both cellular DNA and proteins. The mechanism of their action is thought to occur via the combination of oxidative damage intervened by redox-cycling, adduct formation with DNA and proteins, and protein cross-linking/protein conformation change.

This chapter was previously published in *Quinones: Occurrence, Medicinal Uses and Physiological Importance*, edited by Ervin R. Price and Smith C. Johnson, New York: Nova Publishers Inc., 2013.

* Email: jisook-kim@utc.edu

In: Biochemistry Research Summaries. Volume 1
Editors: James T. Walker and Catherine M. Jones
ISBN: 978-1-53610-895-8

Chapter 48

QUINONE MONOACETAL COMPOUNDS IN APPLICATION TO CONTROLLED REACTIONS WITH NUCLEOPHILES[#]

Toshifumi Dohi and Yasuyuki Kita*
College of Pharmaceutical Sciences, Ritsumeikan University, Kusatsu, Shiga, Japan

RESEARCH SUMMARY

A summary of the preparation, synthetic utility, and application of quinone monoacetals is presented with focus on the following points. Quinone monoacetals (QMAs), the oxidized compounds of phenols as well as the desymmetrized alternative of quinones, have attracted considerable interest due to their broad utilities in organic transformations as intermediates and important building blocks for the synthesis of natural products. Recently, increasing interest in the development and utilization of QMAs has been occurring due to their unique bifunctionalities of both α,β-unsaturated carbonyl and allylacetal moieties. The varied reactivities in nucleophilic attack on QMA carbons can occur, for instance, addition to the carbonyl carbon and conjugated addition to the enone moiety. In contrast to these established addition chemistries, the reports of the utility of QMAs in substitution reactions are quite limited. This chapter principally deals with the progress in the emerging theme of the selectivity during the reactions of QMAs toward nucleophiles, especially with emphasis on the latter topic, the section of which starts for i) efficient preparation of QMAs, ii) general guideline for the reactivities of QMAs toward nucleophiles, and iii) newly developed methods for the regioselective introduction of aromatic or alkene nucleophiles by controlled coupling strategies using specific acid catalysts. In particular, our new strategies can now provide attractive synthetic routes to the valuable oxygenated biaryls, terphenyls, dihydrobenzofurans, and other related functionalized compounds. Several important results, such as the syntheses

---

This chapter was previously published in *Quinones: Occurrence, Medicinal Uses and Physiological Importance*, edited by Ervin R. Price and Smith C. Johnson, New York: Nova Publishers Inc., 2013.

* Tel and Fax +81(77)68798229; E-mail: kita@ph.ritsumei.ac.jp

of key modules of natural products and preparation of regio-controlled phenol oligomers, are also discussed for the promising expansion of these future applications.

In: Biochemistry Research Summaries. Volume 1
Editors: James T. Walker and Catherine M. Jones

ISBN: 978-1-53610-895-8

Chapter 49

CATECHOLQUINONES AS SUBSTRATES OF THE NRH: QUINONE OXIDOREDUCTASE 2 IN THE BRAIN AND RETINA[#]

Lucia de Fatima Sobral Sampaio[*]
Laboratory of Biochemistry of Central Nervous System Development.
Biological Sciences Institute, Federal do Pará University. Belém, PA, Brazil

RESEARCH SUMMARY

Quinones are highly toxic products of the degradation of many compounds surging from live organisms. Certain of these highly toxic products are substrates of the NRH: quinone oxidoreductase 2 (NQO2). This flavoenzyme has a ping-pong by catalytic mechanism, where the coenzyme FAD is reduced by rare cosubstrates, such as N-rybosil-hidronicotinamide (NRH) and N-metyl-hidronicotinamide (NMH). The NQO2 detoxifying activity occurs synchronically with the activation of the anti-cancer protein p53, which is primarily activated in response to xenobiotic and radiation. It is not clear if the over activation of the NQO2 produces ionic reactive compounds that are capable of activating p53, or if the xenobiotic presence is capable of triggering a NQO2-p53 binding, which, in turn, activates protein p53. Among the diversity of putative NQO2 substrates, we highlighted catechol quinones produced from catecholamines in the brain and retina. The catecholamines' participation in neurological diseases, as well the NQO2 influence in the neurological diseases' physiopathology, is incontestable. Accordingly, in this chapter, we aim to discuss the implication of the characteristic catechol quinone reductase NQO2 function in the catechol quinones metabolism, which takes place similarly in neurons from the brain and retina, associating with NQO2 cancer-preventing activity and with those neurological diseases related to catecholamine metabolism's dysfunctions.

[#] This chapter was previously published in *Quinones: Occurrence, Medicinal Uses and Physiological Importance*, edited by Ervin R. Price and Smith C. Johnson, New York: Nova Publishers Inc., 2013.

[*] Email: lsampaio@ufpa.br.

In: Biochemistry Research Summaries. Volume 1
Editors: James T. Walker and Catherine M. Jones
ISBN: 978-1-53610-895-8

Chapter 50

PLASMA MEMBRANE COQ, PORIN, AND REDOX CONTROL OF AUTISM[#]

Brian F. Teske[1], I. L. Sun[2], Anna Gvozdjakova[3], Hans Low[4] and Frederick L. Crane[*,2]

[1]Department of Biochemistry and Molecular Biology, Indiana University, School of Medicine, Indianapolis, IN, US
[2]Department of Biological Science, Purdue University, W. Lafayette, IN, US
[3]School of Medicine, Comenius University, Bratislava, Slovakia
[4]Department of Molecular Medicine and Surgery, Karolinska Institute, Stockholm, Sweden

RESEARCH SUMMARY

Autism is a neurological condition starting in childhood that is characterized by behavioral and intellectual problems. Its occurrence is increasing and although there are some treatments, they are of limited effect or have undesirable side effects. A recent study showed that autistic children had increased serum levels of auto-antibodies to Voltage Dependent Anion Channel (VDAC). Interestingly, in addition to the membrane transport function of VDAC a second function was recently described by A. Lawen's group at Monash University in Melbourne. This group showed that VDAC was also a trans-PM NADH dehydrogenase. The VDAC autoantibody detected in autistic children inhibits the dual transport and dehydrogenase functionality of VDAC. In this report we implicate Coenzyme Q as an important co-factor for redox control of PM pores including VDAC. We show that the PM redox function is dependent on Coenzyme Q and propose that this novel function for CoQ has therapeutic implications for treatment of autism disorders. More broadly, the Coenzyme Q requirement for the PM redox function of porin in diverse species including bacteria, plants, and mammals suggests a mechanistically conserved feature of pore redox control.

[#] This chapter was previously published in *Quinones: Occurrence, Medicinal Uses and Physiological Importance*, edited by Ervin R. Price and Smith C. Johnson, New York: Nova Publishers Inc., 2013.

[*] Address Correspondence to: Frederick L. Crane, flccoq10@aol.com

In: Biochemistry Research Summaries. Volume 1
Editors: James T. Walker and Catherine M. Jones
ISBN: 978-1-53610-895-8

Chapter 51

Evolutionary Aspects of Oleic Acid Biosynthesis and Incorporation from Unicellular to Multicellular Organisms[#]

Cínthia C. Bonatto[1] and Luciano P. Silva[2,*]
[1]Embrapa Genetic Resources and Biotechnology, Brasilia, Brazil
[2]University of Brasilia, Institute of Biological Sciences, Brasilia, Brazil

Research Summary

The idea of life's origin in the context of lipid membranes and fatty acids can be considered uncontroversial. Considering the unique physical-chemical properties of oleic acid, it is not difficult to understand its biological importance. Despite its cytotoxicity against several cell types, there are powerful health benefits and biotechnological applications of oleic acid. Therefore, it is not surprisingly that oleic acid is ubiquitous in nature. A systematic literature compilation has been undertaken on the oleic acid relative content of more than one thousand of living organism species belonging to all the major taxonomic groups. In comparison with other C18 fatty acids, oleic acid is the most abundant type among all biological Domains, but among biological Kingdoms, Plantae is the unique in which oleic acid is not the most abundant C18 fatty acid. The simultaneous use of oleic acid and other C18 fatty acids' content was sufficient to distinguish and hierarchically cluster main taxonomic groups in accordance to common and reliable phylogenetic approaches. In light of these facts, the use of oleic acid and other C18 fatty acids' content could effectively benefit taxonomic inferences and even represents a novel outlook of biological diversity and evolution.

[#] This chapter was previously published in *Oleic Acid: Dietary Sources, Functions and Health Benefits*, edited by Luciano Paulino Silva, New York: Nova Publishers Inc., 2013.

[*] E-mail address: luciano.paulino@embrapa.br, lucianopaulinosilva@gmail.com

In: Biochemistry Research Summaries. Volume 1
Editors: James T. Walker and Catherine M. Jones
ISBN: 978-1-53610-895-8

Chapter 52

OLEIC ACID: NATURAL VARIATION AND POTENTIAL ENHANCEMENT IN OILSEED CROPS[#]

Noelle A. Barkley and Ming Li Wang*
USDA-ARS, PGRCU, Griffin, Georgia, US

RESEARCH SUMMARY

Oleic acid is a monounsaturated omega 9 fatty acid (MUFA, C18:1) which can be found in various plant lipids and animal fats. Unlike omega 3 (linolenic acid, C18:3) and omega 6 (linoleic acid, C18:2) fatty acids which are essential because they cannot be synthesized by humans and must be obtained from dietary sources, omega 9 is not an essential fatty acid and can be synthesized de novo.

Even though oleic acid is not essential, it plays important biological roles in both plants and animals. Consumption of high oleic oils leads to a wide range of beneficial effects on promoting human health (from enhancing immune responses to reducing risks of cardiovascular diseases). Plant oils are one of the most important dietary sources of oleic acid.

The level of oleic acid in plant oils is one of the key factors to determine their nutritional quality and total shelf life. Significant variability in the levels of oleic acid has been identified in different plant oils. Natural variation of oleic acid levels also exists in the oil extracted from the same species likely due to genotype, quantitative trait loci, or environmental effects.

Plant breeders have been using traditional breeding methods to enhance the level of oleic acid in oilseed crops, but the level of enhancement is limited. Genetic modification (including chemical mutagenesis, physical radiation, random bombardment, target-specific modification) has played a big role in altering the total fatty acid composition and has been successfully used on oilseed crops (including soybean, canola, sunflower, cotton, and peanut) for enhancing oleic acid levels. Here in this chapter, we review the known biological function of

---

[#] This chapter was previously published in *Oleic Acid: Dietary Sources, Functions and Health Benefits*, edited by Luciano Paulino Silva, New York: Nova Publishers Inc., 2013.

* E-mail address: MingLi.Wang@ars.usda.gov.

oleic acid, the beneficial effects on human health, dietary sources, and natural variation and potential enhancement on the level of oleic acid in crop plants.

In: Biochemistry Research Summaries. Volume 1
Editors: James T. Walker and Catherine M. Jones

ISBN: 978-1-53610-895-8

Chapter 53

VOLUME AND ADSORPTION PROPERTIES OF OLEIC ACID SOLUTIONS[#]

Victor V. Korolev*, Anna G. Ramazanova, Olga V. Balmasova, Evgenii V. Ivanov and Valentina I. Yashkova

G.A. Krestov Institute of Solution Chemistry of the Russian Academy of Science, Ivanovo, Russia

Nanomagnetic Co., Ltd., Ivanovo, Russia

RESEARCH SUMMARY

The present chapter covers the questions of experimental study of volume and adsorption properties of oleic acid on the surface of superfine magnetite. Oleic acid is the most common unsaturated fatty acid in nature; however considering of its properties is more interesting in comparison with its nearest homologous analogs (linoleic and linolenic acids) exhibiting intrinsic differences due to double bonds number. The limiting partial molar volumes of the solutes and excess molar volumes of their mixtures with nonpolar solvents were discussed. The above parameters were calculated from experimental data on density of dilute solutions of oleic, linoleic, and linolenic acids in tetrachloromethane, n-hexane, n-heptane, cyclohexane, and benzene at 298.15 K. High-precision vibration densimetry method was applied for density measurements; the measurements error did not exceed 5 10^{-6} g/cm^3. It was found that the processes of self-association of the mixtures components played the key role in dissolution effects of fatty acids in CCl_4. The increasing of C=C double bonds number in the fatty acid molecules make the acid desolvation less pronounced and was attended by compression (reduction of volume) of the structure packing of the solvatocomplex formed due to the strengthening interaction of the fatty acid with tetrachloromethane. It was shown that the solutions of fatty acids in *n*-alkanes became closer packed on going from oleic to

[#] This chapter was previously published in *Oleic Acid: Dietary Sources, Functions and Health Benefits*, edited by Luciano Paulino Silva, New York: Nova Publishers Inc., 2013.

* Viktor V. Korolev, e-mail address: vvk@isc-ras.ru.

linolenic acid and with decreasing in hydrocarbon chain length in the solvent molecules. This is caused, as a rule, by the enhancement of the *n*-alkane—acid intermolecular interaction. It was established that benzene influenced selectively on the volumetric effects of dissolution of linoleic and linolenic polyunsaturated fatty acids. The adsorption properties of oleic acid and its analogs on the superfine magnetite surface were studied by equilibrium adsorption method. The sorption - desorption isotherms of oleic, linoleic or linolenic acids from their solutions in carbon tetrachloride, hexane, heptane or cyclohexane on the magnetite surface were obtained. The influence of the surfactants and the solvents on the adsorption properties was discussed. The Theory of Volume Filling of Micropores (TVFM) was applied to calculate the adsorption isotherms parameters, such as limiting adsorption, adsorption characteristic energy, and porous space volume, and their changes depending on surfactant and solvent nature were considered. The two-dimensional state diagrams of adsorption layers of the fatty acids on the solid surfaces were performed.

In: Biochemistry Research Summaries. Volume 1
Editors: James T. Walker and Catherine M. Jones
ISBN: 978-1-53610-895-8

Chapter 54

MICROBIAL TRANSFORMATIONS OF OLEIC ACID[#]

Fabiano J. Contesini[*], José V. Madeira Jr., Danielle B. Lopes, Paula Speranza, Elaine B. Ceresino, Camilo B. Teixeira and Gabriela A. Macedo
Department of Food Science, Faculty of Food Engineering, University of Campinas (UNICAMP), Campinas, Brazil

RESEARCH SUMMARY

Oleic acid is one of the most important vegetable oils fatty acids within different industrial fields. This monounsaturated fatty acid is extensively found in olive, corn, and soybean oils, among others. Its importance varies from health-promoting properties, such as the prevention of cardiovascular diseases and some types of cancer, to many technological applications, taking into account the oxidative stability of this oil. One important application of oleic acid is its use for obtaining different compounds through microbial transformations and biocatalytic processes. These steps include hydration or epoxidation of the double bond, reduction of carboxylic acid and combinations of these reactions. An important group that can be obtained through transformation of oleic acid is the hydroxy fatty acids (HFA) and their derivatives. They are used in cosmetics, paints and coatings, lubricants and in the food industry. Several microorganisms, which include bacterial and yeast strains, produce HFA from oleic acid, such as in the obtainment of (E) 10-hydroxy-8-octadecenoic acid, using a *Pseudomonas* sp. 42A2 strain. 10-hydroxyoctadecanoic acid can be produced by *Nocardia cholesterolicum*, *Rhodococcus* sp. and *Saccharomyces cerevisiae*. These compounds are examples of monohydroxy derivatives of oleic acid and have a diverse potential for application in the food industry, considering its similarity with ricinoleic acid. Added to the possibility of biotransformation of oleic acid using microbial cells, the use of isolated lipases for the production of other compounds from oleic acid and/or triacylglycerols containing oleic

---

[#] This chapter was previously published in *Oleic Acid: Dietary Sources, Functions and Health Benefits*, edited by Luciano Paulino Silva, New York: Nova Publishers Inc., 2013.

[*] E-mail address: fabiano.contesini@gmail.com.

acid have been reported, as well as the production of structured lipids by acydolysis of different oil sources with free fatty acid mixtures. These reactions are catalyzed by microbial lipases from different microorganisms like *Rhizomucor miehei*, and *Thermomyces lanuginosus*. Therefore, this review reports the use of free oleic acid and/or different oils containing oleic acid in their compositions for the obtaining of different important compounds, with the use of microbial cells by biotransformation of oleic acid, or by biocatalytic reactions using isolated microbial lipases.

In: Biochemistry Research Summaries. Volume 1
Editors: James T. Walker and Catherine M. Jones
ISBN: 978-1-53610-895-8

Chapter 55

OLEIC ACID: CONVERSION OF FREE FATTY ACIDS INTO BIODIESEL OVER HETEROGENEOUS CATALYSTS[#]

Catarina S. Caetano and José E. Castanheiro*
Centro de Química de Évora, Departamento de Química,
Universidade de Évora, Portugal

RESEARCH SUMMARY

Biodiesel, which is recognised as a "green fuel", is an alternative to petroleum diesel due to its lower dependence on foreign crude oil and lower combustion emission profile (especially SO2). It is biodegradable and non-toxic, so it is beneficial from an environmental viewpoint. Biodiesel is generally synthesised by transesterification and esterification processes from vegetable oils, animal fats and wastes, which are carried out by acid or base catalysis.

The esterification reaction of long chain carboxylic acids, such as oleic acid, is interesting in the context of biodiesel production. In this process, acid catalysts are more efficient than base catalysts, which have led to the search for more active catalysts for both reactions.

Traditionally, these processes are carried out over homogenous catalysts. However, to make biodiesel production a "Green Process", homogenous catalysts have been replaced with heterogeneous ones. Heterogeneous catalysts have thus recently been used in esterification and transesterification reactions for biodiesel synthesis. Different solid acid catalysts, such as zeolites, resins, tungstated and sulphated zirconia, heteropolyacid, and ionic liquids, have been explored as heterogeneous catalysts for biodiesel production. In this chapter, the use of different solid catalysts for the conversion of free fatty acids (especially oleic acid) into biodiesel will be reviewed.

[#] This chapter was previously published in *Oleic Acid: Dietary Sources, Functions and Health Benefits*, edited by Luciano Paulino Silva, New York: Nova Publishers Inc., 2013.

[*] E-mail address: jefc@uevora.pt

In: Biochemistry Research Summaries. Volume 1
Editors: James T. Walker and Catherine M. Jones

ISBN: 978-1-53610-895-8

Chapter 56

SOURCES OF OLEIC ACID AND THEIR TECHNOLOGICAL AND HEALTH IMPLICATIONS

Rouxléne van der Merwe[1,*] and Arno Hugo[2]
[1]Department of Plant Sciences, University of the Free State, South Africa
[2]Department of Microbial, Biochemical and Food Biotechnology, University of the Free State, South Africa

RESEARCH SUMMARY

Increased concern about health risks of foods has led to demands for improvement in the quality control of vegetable oils. Oil quality refers not only to the nutritional properties, but also the functionality of the oil. Optimal oil quality does not exist, as it will depend on the end-use of the oil.

The fatty acid composition is one of the main parameters defining oil nutritional quality. The fatty acid composition of edible oil plays an important role in the shelf life, nutritional, and health properties of the oil. Polyunsaturated fatty acids have been desirable for their role in lowering plasma cholesterol levels. High amounts of polyunsaturated fatty acids are present in edible oils such as soybean, safflower, sunflower and peanut.

However, polyunsaturated oils are susceptible to oxidative deterioration and consequently need additional processing and chemical additives that result in a final product with detrimental implications for human health. Monounsaturated fatty acids and especially oleic acid are naturally stable to oxidation and have good frying stability and shelf-life properties.

In fact, oils rich in oleic acid have positive health implications because of their low saturated fatty acid contents and low *trans*-isomer levels. They offer the consumer the choice of better quality oil with reduced health risks.

* Email address: vandermerwer@ufs.ac.za

In: Biochemistry Research Summaries. Volume 1
Editors: James T. Walker and Catherine M. Jones
ISBN: 978-1-53610-895-8

Chapter 57

OLEIC ACID IN MILK OF DIFFERENT MAMMALIAN SPECIES#

Paolo Polidori[1]* and Silvia Vincenzetti[2]
[1]School of Pharmacy, University of Camerino, Matelica (MC), Italy
[2]School of Veterinary Medical Sciences, University of Camerino, Via Circonvallazione, Matelica (MC), Italy

RESEARCH SUMMARY

Milk, being a biological fluid, varies considerably from different mammal species in several of its constituents. Fat content of milk is the nutrient showing the most variation in its percentage. In fact, milk fat is one of the most complex of all common fats, composed of about 98% triglycerides; other milk lipids (as a percentage of the total milk fat) include diacylglycerides (0.25-0.48%), monoacylglycerides (0.02-0.04%), phospholipids (0.6-1.0%), cholesterol (0.2-0.4%), glycolipids (0.006%) and free fatty acids (0.1-0.4%). Milk fat contains many fatty acids, mainly saturated but also monounsaturated and polyunsaturated. Several factors, many of which are interactive, are known to cause variation in milk fatty acid composition. These include stage of lactation, seasonal variations, feeding and animal genetics. Milk's fatty acid composition has been frequently investigated because of its implications for human health; due to the relationship between high-fat diets and heart diseases, consumer interest in milk fat content and also in fatty acid composition of foods has grown in the recent years. Oleic acid (C18:1cis9), a monounsaturated fatty acid, has been deeply studied because of its possible beneficial effects on human health, normally associated with a reduction of serum cholesterol and plasma cholesterol levels. Scientific evidence and nutritional guidelines recommend a reduction in total fat intake, and particularly of saturated fatty acids, which are associated with an increased risk of obesity, hypercholesterolemia and some cancers. In mammalian species, milk fat has a double origin: circulating fatty acids are

---

This chapter was previously published in *Oleic Acid: Dietary Sources, Functions and Health Benefits*, edited by Luciano Paulino Silva, New York: Nova Publishers Inc., 2013.

* E-mail address: paolo.polidori@unicam.it.

transported in lipoproteins, other fatty acids are synthesized in the udder. The fatty acid composition of milk fat is of interest because it affects both the texture and the flavor of dairy products. Milk fat in ruminant species is characterized by a higher content of saturated fatty acids, whereas fatty acid composition in mare's milk and donkey's milk shows interesting percentages of essential fatty acids. Oleic acid is the most abundant MUFA in all the different milk samples considered in this chapter, and its importance to prevent cardiovascular diseases and some inflammatory disorders is confirmed by different clinical studies.

In: Biochemistry Research Summaries. Volume 1
Editors: James T. Walker and Catherine M. Jones
ISBN: 978-1-53610-895-8

Chapter 58

OLEIC ACID AND BLOOD BRAIN BARRIER[#]

Hyung Soo Han*
Department of Physiology, Brain Science and Engineering Institute,
Kyungpook National University School of Medicine, Daegu, Korea

RESEARCH SUMMARY

Blood brain barrier (BBB) acts as a physical and functional barrier between blood circulation and brain parenchyma. Since BBB maintains the cerebral microenvironment in a strict manner, it is essential in protecting the brain from the exogenous deleterious substances such as bacteria, viruses, or toxins in the pathological conditions. But tight restriction of BBB poses a significant obstacle to the delivery of most drugs into the brain. As a result, drug therapy has rarely been available to treat brain tumors, cancer metastases to brain and other brain diseases and therefore delivery system to overcome BBB is crucial to the central nervous system (CNS) drug developers. Fatty acids are the major constituents of lipids in our body. The nature of the fatty acids in the cell membrane contributes greatly to the fluidity of the membrane and it is believed that fatty acid plays a role in regulating the activity of membrane proteins. In addition, membrane phospholipids are a source of second messenger molecules, such as diacylglycerol, phosphatidic acid, inositol 1,4,5-triphosphate, ceramide and arachidonic acid that are involved in the regulation of signaling pathways from the membrane to the cytosol or nucleus. Some fatty acids are substrates for the synthesis of bioactive molecules such as prostaglandins, thomboxanes and leukotrienes. Oleic acid is one of the major constituents of unsaturated fatty acids of human fat. In experimental model of cerebral fat embolism induced by intra-arterially administered oleic acid, brain edema appeared in the hyperacute stage. This transient edema formation is related with reversible dysfunction of BBB. The mechanisms of enhanced BBB permeability by oleic acid have been investigated. Oleic acid induced BBB alteration was visualized by microscopic observation of extravasated Evans blue or FITC-labeled dextran in addition to gadolinium-enhanced T1

[#] This chapter was previously published in *Oleic Acid: Dietary Sources, Functions and Health Benefits*, edited by Luciano Paulino Silva, New York: Nova Publishers Inc., 2013.

* E-mail address: hshan@knu.ac.kr

magnetic resonance imaging (MRI). Dysfunction of BBB was quantified by measuring Evans blue concentration in the brain tissue or counting endothelial barrier antigen (EBA) immunoreactivity. Oleic acid induced inflammation in the cerebral vessels and surrounding brain tissues was observed. Induction of intercellular adhesion molecule-1 (ICAM-1) in the vascular endothelial cells and migration of neutrophils into the brain parenchyma were observed. In conclusion, inflammation might be one of the major contributing factors of oleic acid induced BBB disruption.

In: Biochemistry Research Summaries. Volume 1
Editors: James T. Walker and Catherine M. Jones
ISBN: 978-1-53610-895-8

Chapter 59

FOOD SUPPLEMENTATION WITH VEGETABLE OILS RICH IN OLEIC ACID: HEALTH-PROMOTING EFFECTS AND APPLICATION VIABILITY[#]

Fabiano J. Contesini[*], Elaine B. Ceresino, Ruann J. S. Castro, Francisco F. C. Barros, Evelyn Sassaki and Hélia H. Sato

Department of Food Science, College of Food Engineering, State University of Campinas (UNICAMP), Campinas, SP, Brazil

RESEARCH SUMMARY

The role of food in human health is broad due to its diverse functions. Many foods may be explored in different aspects, since they are not only important for basic nutrition, but also show interesting functional properties. Unsaturated oils are commonly used in food supplementation, taking into account that they are sources of fatty acids, such as oleic acid. The regular consumption of these foods is associated with healthy effects on the human body, including reduction of coronary heart disease risk and beneficial effects on blood lipids, including reduction of total and LDL-cholesterol, triglycerides and increase of HDL-cholesterol in blood. The substitution of saturated for unsaturated fatty acids, as well as the supplementation with vegetable oils rich in oleic acid have been used in food technology as a strategy of food production, with improvement in functional and sensory characteristics. Milk is an example of these substitutions, widely applied in the food industry. Various research studies have shown that the consumption of milk enriched with oleic acid promoted a myriad of positive effects in healthy volunteers, subjects with increased risk factors and cardiovascular patients. The effects of supplementation with oleic acid in different types of salami and other products of animal origin have also been reported. Wheat breads

---

[#] This chapter was previously published in *Oleic Acid: Dietary Sources, Functions and Health Benefits*, edited by Luciano Paulino Silva, New York: Nova Publishers Inc., 2013.

[*] E-mail address: fabiano.contesini@gmail.com.

supplemented with oleic-rich sunflower seed were produced without any significant adverse effects regarding the crust color, crumb grain structure and uniformity. Oleic acid has also been used in edible coatings such as plasticizer and emulsifier, for example, to preserve the quality of cold-stored strawberries, carrots and other fresh products. They improve sample appearance and confer significant property barriers. These are important sensory and technological characteristics that justify the process viability, besides the addition of high nutritional value. Therefore, food supplementation with vegetable oils rich in oleic acid can be very interesting, and different studies about this topic are going to be reviewed in this manuscript.

In: Biochemistry Research Summaries. Volume 1
Editors: James T. Walker and Catherine M. Jones
ISBN: 978-1-53610-895-8

Chapter 60

APPLICATIONS OF OLEIC ACID ON NANOTECHNOLOGY CONCERNING ENZYME IMMOBILIZATION TECHINIQUES[#]

Eduardo F. Barbosa*[*] *and José L. Cardozo Filho
Institute of Biology, University of Brasilia, Brazil
Embrapa Genetic Resources and Biotechnology, Brasilia, Brazil

RESEARCH SUMMARY

This chapter describes a review about oleic acid applications in magnetic systems structured at billionth part of matter, related to enzymes immobilization. Basically, it is shown the oleic acid stabilizing effects over magnetic nanoparticles, mainly those derived from iron. Some oleic acid properties are discussed specifically concerning the reducing effects in magnetic nanoparticles magnetization saturation, as well as size control improvements. Protocols for nanostructures production were gathered to provide compact information about different routes to obtain controllably nanostructured systems, allowing the reader to compare some synthesis strategies. A compilation of studies related to enzyme immobilization (mainly lipases from different organisms) using oleic acid based materials was presented, highlighting the benefits obtained from this technique, such as a better tolerance against adverse catalysis conditions and kinetic parameters. Other oleic acid based materials applications were described, showing the high versatility of these materials. Finally, this chapter points out the use of nanostructured oleic acid based systems as support for enzyme immobilization like an opened research field, which requires extensive investigations to whole clarify its application and potentialities on biotechnology.

[#] This chapter was previously published in *Oleic Acid: Dietary Sources, Functions and Health Benefits*, edited by Luciano Paulino Silva, New York: Nova Publishers Inc., 2013.
[*] E-mail address: barbosaeduardofernandes@gmail.com

In: Biochemistry Research Summaries. Volume 1
Editors: James T. Walker and Catherine M. Jones

ISBN: 978-1-53610-895-8

Chapter 61

Oleic Acid and Cancer: From Therapeutic to Co-Adjuvant Agents in Anticancer Approaches[#]

Graziella A. Joanitti[*]
Institute of Biological Sciences, University of Brasilia, Brazil
Embrapa Genetic Resources and Biotechnology, Brasilia, Brazil

Research Summary

The heterogeneity and versatility nature of cancer emphasize why cancer treatment is considered a challenging task. The search for new tools has not only been based on the discovery of new molecules, but also on the discovery of new functions for known molecules. Oleic acid is one of those examples. Oleic acid is a monounsaturated (C18:1, n-9) fatty acid found in animal and plant fats often included in the human diet and is already known for its nutritional and structural properties. However, this fatty acid has also been extensively studied for its anticancer properties. The aim of this chapter is to gather the current knowledge regarding the effects of oleic acid in cancer and to provide a review of its versatile applications as an anticancer molecule itself, a co-factor/stabilizing agent, a skin penetration enhancer molecule, and as nanoparticle components aiming to improve, complement, and enable developing anticancer therapies.

[#] This chapter was previously published in *Oleic Acid: Dietary Sources, Functions and Health Benefits*, edited by Luciano Paulino Silva, New York: Nova Publishers Inc., 2013.

[*] E-mail address: bygra1@gmail.com

In: Biochemistry Research Summaries. Volume 1
Editors: James T. Walker and Catherine M. Jones
ISBN: 978-1-53610-895-8

Chapter 62

BIOLOGICAL ACTIVITIES OF KAEMPFEROL: EFFECT OF CYCLODEXTRINS COMPLEXATION ON THE PROPERTIES OF KAEMPFEROL[#]

M. T. Mercader-Ros, C. Lucas-Abellán, M. I. Fortea, A. Serrano-Martínez, J. A. Gabaldón and E. Núñez-Delicado
Department of Food Technology and Nutrition,
San Antonio Catholic University, Guadalupe. Murcia, Spain

RESEARCH SUMMARY

The flavonoid kaempferol is a yellow compound with a low molecular weight (MW: 286.24 g/mol) that is commonly found in plant-derived foods and in plants used in traditional medicine. Because the enzymes involved in the biosynthesis of kaempferol are relatively common in the plant kingdom, it is not surprising that this flavonoid is widely distributed in plants. Kaempferol is relatively abundant in apples, grapes, gingko biloba, onions, leeks, citrus fruits, red wines, tea, cruciferous vegetables, and some edible berries. Many preclinical studies have shown that kaempferol and some glycosides of kaempferol have extensive pharmacological activities, including antioxidant, anti-inflammatory, anticancer, antimicrobial, cardioprotective, neuroprotective, antidiabetic, anti-osteoporotic, estrogenic/antiestrogenic, anxiolytic, analgesic and antiallergic activities. The above-mentioned antioxidant properties reside mainly in their radical-scavenging activity. Repeated studies have shown that flavonoids with many hydroxyl groups are extremely effective antioxidants. In fact, hydroxylated flavonols, such as myricetin, quercetin and kaempferol, have been demonstrated to be particularly effective antioxidants in many studies.

[#] This chapter was previously published in *Kaempferol: Chemistry, Natural Occurrences and Health Benefits*, edited by Garion Villers and Yves Fougere, New York: Nova Publishers Inc., 2013.

All these beneficial health effects attributed to kaempferol, make this natural compound interesting for the development of functional foods. The main problem is its low water solubility and its stability against various factors such as oxygen, light and temperature.

In recent years, complexation with cyclodextrins (CDs) has successfully been used to improve the solubility, chemical stability and bioavailability of a number of poorly soluble phenolics compounds, such as resveratrol, mirycetin and quercetin. CDs are a group of naturally occurring cyclic oligosaccharides derived from starch with six (α-), seven (β-) or eight (γ-cyclodextrins) glucose residues, linked by $\alpha(1 \rightarrow 4)$ glycosidic bonds. In the pharmaceutical, cosmetics and food industries, CDs have been used as complexing agents to increase the water solubility of various compounds, such as drugs, vitamins and food colourants, increasing the solubility, stability and bioavailability of the guest molecule. Kaempferol can be included in CDs increasing the solubility, possibly its bioavailability and beneficial health effects.

In: Biochemistry Research Summaries. Volume 1
Editors: James T. Walker and Catherine M. Jones
ISBN: 978-1-53610-895-8

Chapter 63

KAEMPFEROL AND KAEMPFERITRIN: NUTRACEUTICAL COMPOUNDS CONTRIBUTE TO GLUCOSE HOMEOSTASIS BY ACTING AT MULTIPLE BIOLOGICAL SITES[#]

***Fátima Regina Mena Barreto Silva*[*,1], *Leila Zanatta*[†,2], *Marisa Jádna Silva Frederico*[‡,1], *Moacir Geraldo Pizzolatti*[§,3] *and Angela Machado De Campos*[‖,4]**

[1]Departamento de Bioquímica, Centro de Ciências Biológicas, Universidade Federal de Santa Catarina, Campus Universitário, Bairro Trindade, Florianópolis, SC, Brazil
[2]Área de Ciências da Saúde, Universidade Comunitária da Região de Chapecó, Chapecó, SC, Brazil
[3]Departamento de Química, Centro de Ciências Físicas e Matemáticas,
[4]Departamento de Química Farmacêutica, Centro de Ciências da Saúde, Universidade Federal de Santa Catarina, Campus Universitário, Florianópolis, SC, Brazil

RESEARCH SUMMARY

Kaempferol is a flavonol that is relatively abundant in various plants and some edible berries. Kaempferol aglycone or its glycosilated forms contributes significantly to flavonoid intake in humans (about 33% from total flavonol intake), with intakes estimated around 10 mg per day in developed countries. These compounds have antioxidative, anti-inflammatory, antihypertensive, lipolytic and anti-cancer effects. Emerging evidence suggests that

---

[#] This chapter was previously published in *Kaempferol: Chemistry, Natural Occurrences and Health Benefits*, edited by Garion Villers and Yves Fougere, New York: Nova Publishers Inc., 2013.

[*] mena@mbox1.ufsc.br.a

[†] leila.zanatta@gmail.com.

[‡] marisafrederico1@hotmail.com.

[§] moacir.pizzolatti@gmail.com.

[‖] angelacampos@gmail.com.

kaempferol exhibits nutraceutical properties since it produces physiological benefits and is able to combat disease processes, among them those concerned to diabetes mellitus. Several results have demonstrated that kaempferol can influence glucose metabolism acting as antihyperglycemic and especially as hypoglycemic agent through *in vitro* and *in vivo* approaches. Kaempferol aglycone and/or glycosides stimulate glucose uptake and glycogen synthesis in skeletal muscle, inhibit intestinal glucose absorption and act *per se* as insulin-mimetic, probably, by influencing the pleiotropic mechanisms of insulin signaling. Thus, these nutraceutical compounds have attracted the interest of both academia and industry as possible therapy options. This review cover the frontiers of the development of chemistry and biological properties of kaempferol (aglycone/glycosides), especially those associated with glucose homeostasis and carbohydrate metabolism able to ameliorate diabetes status. Moreover, updates the debate about the development of new approaches to study the interactions of nutraceutical compounds with target-sites of action combining *in vivo, ex vivo* and *in vitro* pharmacological studies. Also, the perspectives of controlled delivering of these substances in multiple and specific biological locations useful to diabetes mellitus therapy are highlighted.

In: Biochemistry Research Summaries. Volume 1
Editors: James T. Walker and Catherine M. Jones
ISBN: 978-1-53610-895-8

Chapter 64

DISTRIBUTION OF KAEMPFEROL GLYCOSIDES AND THEIR FUNCTION IN PLANTS[#]

Tsukasa Iwashina[1,*] and Yoshinori Murai[2]
[1]Department of Botany, National Museum of Nature and Scieince, Tsukuba, Japan
[2]Department of Chemistry, Hiyoshi Campus, Keio University, Yokohama, Japan

RESEARCH SUMMARY

About 350 kaempferol glycosides are found in plant kingdom. Their glycosides are distributed to most plant species from mosses and ferns to vascular plants. Though the flavonoids involving kaempferol were considered to be waste products of plant metabolism in the early days of 20th century, they now may be to serve as many functions and activities in land plants. Copigmentation with the anthocyanins was well-known as a function of the kaempferol glycosides in the flowers, especially blue and purple flowers. Additionally, it is known that kaempferol, which can hardly be seen by human eyes, also act as the important pollinator attractants in addition to visible anthocyanins. Recently, other functions, e.g., feeding attractants (kaempferol 3-*O*-xylosyl-(1→2)-galactoside), allelopathy (kaempferol itself), phytoalexins (kaempferol 3-*O*-robinobioside-7-*O*-rhamnoside) of naturally occurring kaempferol, were reported by some authors. In this review, we describe the distribution of kaempferol glycosides and their functions and activities in plants.

[#] This chapter was previously published in *Kaempferol: Chemistry, Natural Occurrences and Health Benefits*, edited by Garion Villers and Yves Fougere, New York: Nova Publishers Inc., 2013.

[*] E-mail: iwashina@kahaku.go.jp.

In: Biochemistry Research Summaries. Volume 1
Editors: James T. Walker and Catherine M. Jones

ISBN: 978-1-53610-895-8

Chapter 65

ANALYTICAL APPROACHES FOR KAEMPFEROL DETERMINATION[#]

Ondrej Biba[1,*], Miroslav Strnad[1,2] and Jiri Gruz[1,2,†]
[1]Centre of the Region Hana for Biotechnological and Agricultural Research, Faculty of Science, Palacky University, Olomouc, Czech Republic
[2]Laboratory of Growth Regulators, Institute of Experimental Botany AS CR, Olomouc, Czech Republic

RESEARCH SUMMARY

The rapid development of technology and analytical instruments in the last 20 years allowed scientists to detect and quantify thousands of compounds which were difficult to analyse before. Kaempferol, a common biologically active phytochemical, occurs mostly in complex matrices together with other plant metabolites that complicate the analytical procedure. Various extraction and purification steps, including solid phase extraction, were therefore developed to facilitate the determination of kaempferol in complex biological samples. This chapter is an overview of currently used analytical instruments and their basic settings used to detect and quantify kaempferol in various plants, plant-derived sources and other biological samples.

[#] This chapter was previously published in *Kaempferol: Chemistry, Natural Occurrences and Health Benefits*, edited by Garion Villers and Yves Fougere, New York: Nova Publishers Inc., 2013.

[*] Ondrej Biba: Centre of the Region Hana for Biotechnological and Agricultural Research, Faculty of Science, Palacky University, Slechtitelu 11, CZ 783 71 Olomouc, Czech Republic.

[†] Corresponding author: Jiri Gruz. Centre of the Region Hana for Biotechnological and Agricultural Research, Faculty of Science, Palacky University, Slechtitelu 11, CZ 783 71 Olomouc, Czech Republic. Laboratory of Growth Regulators, Institute of Experimental Botany AS CR, Slechtitelu 11, CZ 783 71 Olomouc, Czech Republic. E-mail: jiri.gruz@ gmail.com.

In: Biochemistry Research Summaries. Volume 1
Editors: James T. Walker and Catherine M. Jones
ISBN: 978-1-53610-895-8

Chapter 66

CHEMOPROTECTIVE EFFECTS OF KAEMPFEROL IN COLORECTAL TUMORIGENESIS#

Gerd Bobe*[*,1,2], Matthew Young[3], Elaine Lanza[3], Amanda J. Cross[4] and Nancy H. Colburn[3]

[1]Linus Pauling Institute, Oregon State University, Corvallis, OR, US
[2]Department of Animal and Rangeland Sciences, Oregon State University, Corvallis, OR, US
[3]Laboratory of Cancer Prevention, Center for Cancer Research, National Cancer Institute, Frederick, MD, US
[4]Nutritional Epidemiology Branch, Division of Cancer Epidemiology and Genetics, National Cancer Institute, National Institutes of Health, Rockville, MD, US

RESEARCH SUMMARY

Kaempferol is a bioactive polyphenol that has been shown to have chemoprotective properties in cell culture. The objective of this study was to evaluate the potential chemoprotective effects of kaempferol against colorectal tumorigenesis in humans and animal models. The association between kaempferol intake and colorectal adenoma recurrence was examined in the Polyp Prevention Trial (PPT), a 4-year randomized, clinical intervention trial which examined the effect of a high-fruit and vegetable, high-fiber, and low-fat diet on colorectal adenoma recurrence. Kaempferol consumption was estimated annually by self-administered food frequency questionnaires that were connected to the U.S. Department of Agriculture flavonoid database for foods.

Kaempferol intake was inversely associated with high-risk adenoma [odds ratio (OR) for individuals in the highest compared to those in the lowest quartile: 0.62; 95% confidence

This chapter was previously published in *Kaempferol: Chemistry, Natural Occurrences and Health Benefits*, edited by Garion Villers and Yves Fougere, New York: Nova Publishers Inc., 2013.

* Correspondence and current affiliation: Gerd Bobe, 307 Linus Pauling Science Center, Oregon State University, Corvallis, OR 97331; e-mail: gerd.bobe@oregonstate.edu; phone: 541-737-1898; fax: 541-737-4174.

interval (CI): 0.40, 0.97; $P_{\text{Linear}} = 0.04$) and advanced adenoma recurrence (OR: 0.49; 95% CI: 0.26, 0.91; $P_{\text{Linear}} = 0.03$).

To examine whether the chemoprotective effect of kaempferol could be in part explained by its anti-inflammatory properties, we measured serum concentrations of cytokines at baseline, year 1, and year 3 in the intervention arm of the PPT. High intake of kaempferol was associated with lower serum interleukin 6 (IL-6) concentrations in year 1 and 3 (highest versus lowest kaempferol intake quartile: 1.86 versus 2.09 pg/mL; $P_{\text{Kaempferol}} = 0.05$). The greatest decrease in serum IL-6 concentrations was observed in participants with elevated serum IL-6 concentrations (> 2.37 pg/mL) at baseline (highest versus lowest intake quartile: 2.63 versus 3.23 pg/mL; $P_{\text{Kaempferol}} = 0.05$). Individuals with above median kaempferol consumption and equal or below median change in IL-6 concentrations had the lowest risk of advanced adenoma recurrence.

To examine whether the beneficial effects of kaempferol on colorectal tumorigenesis were associated with other compounds enriched in kaempferol-containing foods, we conducted a study in male *FVB/NCr* mice fed a diet with or without 500 ppm purified kaempferol for 85 days after colorectal tumor induction and promotion by the chemical carcinogen azoxymethane and the colonic irritant dextran sodium sulfate. In comparison to mice on the control diet, mice consuming kaempferol had fewer tumors (16 vs. 21 tumors; $P = 0.04$) and smaller tumors (11.6 vs. 7.4 mm^3/tumor; $P = 0.09$).

Our results suggest kaempferol is a potential chemoprotective compound against colorectal tumorigenesis.

In: Biochemistry Research Summaries. Volume 1
Editors: James T. Walker and Catherine M. Jones
ISBN: 978-1-53610-895-8

Chapter 67

MULTIPOTENT FLAVONOID KAEMPFEROL: MOLECULAR TARGETS AND MECHANISM OF ACTION AND NANOTECHNOLOGY APPLICATIONS IN CANCER AND HUMAN HEALTH[#]

Allen Y. Chen[1] and Yi Charlie Chen*[2]
[1]Department of Pharmaceutical Sciences, West Virginia University, Morgantown, WV, US
[2]Natural Science Division, Alderson-Broaddus College, Philippi, WV, US

RESEARCH SUMMARY

Kaempferol, a bioactive natural compound widely present in fruits and vegetables, is a polyphenol belonging to the class of flavonoids. Studies for the past several decades have identified the great potential of this flavonoid in the prevention and treatment of human chronic diseases, especially cancer. Kaempferol exerts this wide spectrum of effects by targeting several proteins/enzymes that are implicated in cancer and disease progression. Epidemiological studies have shown an inverse relationship between kaempferol intake and cancer. Kaempferol helps by augmenting the body's antioxidant defence against free radicals, which promote the development of cancer. At the molecular level, kaempferol has been reported to modulate a number of key elements in cellular signal transduction pathways linked to apoptosis, angiogenesis, inflammation, and metastasis. Significantly, kaempferol inhibits cancer cell growth and angiognesis and induces cancer cell apoptosis, while also appearing to preserve normal cell viability, in some cases exerting a protective effect. Because of its outstanding safety profile and its availability in highly purified forms, the clinical implications for kaempferol are growing rapidly. Even though kaempferol is one of

---

[#] This chapter was previously published in *Kaempferol: Chemistry, Natural Occurrences and Health Benefits*, edited by Garion Villers and Yves Fougere, New York: Nova Publishers Inc., 2013.

[*] Corresponding author: Yi Charlie Chen, Alderson-Broaddus College, 101 College Hill Drive, Box 2096, Philippi, WV 26416, Email: chenyc@ab.edu, Tel: 1-304-457-6277, Fax: 1-304-457-6239.

the major flavonoids in the human diet, its bioavailability has been debated. Recent studies to improve its bioavailability by encapsulating it with PEO-PPO-PEO and PLGA nanoparticles are promising. This review focuses mainly on the molecular targets of kaempferol and the mechanisms by which kaempferol exerts its biological effects in the prevention and treatment of human chronic diseases and cancer. A discussion on the application of nanotechnology is also included.

In: Biochemistry Research Summaries. Volume 1
Editors: James T. Walker and Catherine M. Jones
ISBN: 978-1-53610-895-8

Chapter 68

METAL COMPLEXES OF KAEMPFEROL AND THEIR SPECIATION IN HUMAN PLASMA[#]

***Andrija Ciric**[*,1], **Ivan Jakovljevic**[1], **Milica Cvijovic**[2], **Milena Jelikic-Stankov**[3] and **Predrag Djurdjevic**[1]*
[1]Faculty of Science, University of Kragujevac, Kragujevac, Serbia
[2]Faculty of Agronomy, University of Kragujevac, Čačak, Serbia
[3]Faculty of Pharmacy, University of Belgrade, Belgrade, Serbia

RESEARCH SUMMARY

Flavonoids are a large class of phenolic compounds which are sub classified as flavones, flavonols, isoflavones, flavanones and catechins, chalcones and anthocyanidins depending on phenyl substituent in the C_2 or C_3 position in *benzo-γ-pyrone* nucleus. Interest in the bioflavonoids is related to their diversity, biological significance as secondary plant metabolites and ecological role, use as chemotaxonomic markers, impact on fruit quality, physiological effects and industrial applications.

The flavonoids are potent antioxidants, free radical scavengers and metal chelators; they inhibit lipid peroxidation and exhibit various physiological activities, including anti-inflammatory, anti-allergic, anti-carcinogenic, antihypertensive and anti-arthritic activities.

It is well known that blood serum contains essential amino acids, 12 essential metal ions at least another 100 ligands as well as numerous low molecular weight complexes. In addition, there are α/β macro globulins such as albumin and transferin weighing up to 100000 Daltons. It has been well demonstrated that metal ions play a large number of important roles in biological systems. Metal ions which are considered essential to human life included calcium, magnesium, manganese, iron, cobalt, copper and zinc. Furthermore man is constantly being challenged by pollutants included the toxic metal ions mercury, cadmium and lead.

[#] This chapter was previously published in *Kaempferol: Chemistry, Natural Occurrences and Health Benefits*, edited by Garion Villers and Yves Fougere, New York: Nova Publishers Inc., 2013.

[*] Corresponding author: andrija@kg.ac.rs, tel: +38134300261, fax: +38134335040.

In: Biochemistry Research Summaries. Volume 1
Editors: James T. Walker and Catherine M. Jones
ISBN: 978-1-53610-895-8

Chapter 69

THE ACTION OF KAEMPFEROL ON ORAL CAVITY CANCER AND AIRWAY SECRETION[#]

Kyung-Su Kim[*]
Department of Otorhinolaryngology and Human Barrier Research Institute,
Yonsei University College of Medicine, Seoul, Korea

RESEARCH SUMMARY

Ginkgo biloba extract has long been used as an anticancer drug and expectorant in Oriental medicine. Because *Ginkgo biloba* extract is a mixed compound comprising kaempferol and other components, it may be preferable to use only kaempferol for treatments due to possible adverse reactions with the other components. Furthermore, the action mechanism and potency of kaempferol compared to *Ginkgo biloba* extract have not been revealed. Therefore, this chapter gives a brief survey of the current knowledge on kaempferol and its potential use as an anti-oral cancer agent and also as an anti-hypersecretory agent.

In previous studies, *Ginkgo biloba* extract has been shown to induce apoptosis in oral cavity cancer cells. In the case of kaempferol, the results of cell proliferation assays on oral cavity cancer cell lines such as SCC-1483, SCC-25, and SCC-QLL1 showed that kaempferol significantly inhibited cellular proliferation. Flow cytometry showed that the anti-proliferative effect was due to increased apoptosis. Kaempferol showed cleavage of poly (ADP-ribose) polymerase (PARP). Therefore, kaempferol effectively induces caspase-3-dependent apoptosis of oral cavity cancer cells and can be considered a possible anti-oral cancer agent.

In addition to its anticancer effects, *Ginkgo biloba* extract suppresses IL-1β-induced *MUC5AC* gene expression in NCI-H292 cells, and therefore, kaempferol may be used as a suppressor of mucin production. The RT-PCR and real-time PCR results showed that kaempferol suppressed *MUC5AC* mRNA expression in a dose-dependent manner, with

---

[#] This chapter was previously published in *Kaempferol: Chemistry, Natural Occurrences and Health Benefits*, edited by Garion Villers and Yves Fougere, New York: Nova Publishers Inc., 2013.

[*] Corresponding author: Professor Kyung-Su Kim, Department of Otorhinolaryngology, Gangnam Severance Hospital, Yonsei University College of Medicine, 211 Eonju-ro, Gangnam-gu, 135-720 Seoul, Korea. Tel: +82 2 2019 3463, Fax: +82 2 3463 4750; E-mail address: ydrhinol@yuhs.ac.

significant inhibition starting at 40 μM (equal concentration to about a 1/12or 1/13 dose of *Ginkgo biloba* extract).

Kaempferol suppressed phosphorylation of ERK and p38 kinases. Therefore, kaempferol may overcome the dose problem of *Ginkgo biloba* extract to play a clinically valuable role in controlling mucin hypersecretion in airway inflammation.

In: Biochemistry Research Summaries. Volume 1
Editors: James T. Walker and Catherine M. Jones

ISBN: 978-1-53610-895-8

Chapter 70

KAEMPFEROL REGULATES INTESTINAL TIGHT JUNCTION PERMEABILITY[#]

Takuya Suzuki*
Department of Biofunctional Science and Technology,
Graduate School of Biosphere Science, Hiroshima University, Japan

RESEARCH SUMMARY

A primary function of gastrointestinal epithelial cells is to provide a physical barrier to the diffusion of pathogens, toxins, and allergens from the external environment into the tissues. Defects in the barrier function are involved in the pathogenesis of various intestinal diseases, such as inflammatory bowel diseases, irritable bowel disease, and food allergies. This intestinal barrier is organized through interactions among several barrier components, including the adhesive mucous gel layer, the mucosal immune system, and intercellular tight junctions (TJs). The TJ is a multi-protein complex, which locates around the apical end of the lateral membrane of polarized epithelial cells, and selectively regulates the paracellular passage of ions, molecules and water. The TJ complex is composed of transmembrane proteins that interact with the perijunctional actin cytoskeleton through intracellular plaque proteins. TJ permeability and assembly are dynamically regulated by various intracellular signaling molecules, such as protein kinase C, mitogen-activated protein kinases, and protein phosphatases. Kaempferol is one of the most abundant flavonoids in human diets and exhibits various biological effects, such as anti-inflammatory and anti-cancer activities. A recent study has demonstrated that kaempferol participates in the regulation of intestinal TJ barrier function. This chapter presents the recent advances in our understanding of the kaempferol-mediated suppressive effect on intestinal TJ permeability with a particular focus on intracellular molecular mechanisms.

[#] This chapter was previously published in *Kaempferol: Chemistry, Natural Occurrences and Health Benefits*, edited by Garion Villers and Yves Fougere, New York: Nova Publishers Inc., 2013.

* Corresponding author: Takuya Suzuki, Ph. D., 1-4-4, Kagamiyama, Higashi-Hiroshima, 739-8528, Japan. TEL: 81-82-424-7984. Fax: 81-82-424-7916; E-mail: takuya@hiroshima-u.ac.jp.

In: Biochemistry Research Summaries. Volume 1
Editors: James T. Walker and Catherine M. Jones
ISBN: 978-1-53610-895-8

Chapter 71

IDENTIFICATION OF KAEMPFEROL IN HISTORICAL TEXTILES: A CASE STUDY ON TEXTILES OF BYZANTINE ORIGIN[#]

Dimitrios Mantzouris[1] and Ioannis Karapanagiotis[2,*]
[1]Ormylia Foundation, Art Diagnosis Center, Ormylia, Greece
[2]University Ecclesiastical Academy of Thessaloniki, Department of Management and Conservation of Ecclesiastical Cultural Heritage Objects, Thessaloniki, Greece

RESEARCH SUMMARY

Our goal is to present, in brief, the historical use of kaempferol as a dyestuff material for textiles. For this reason we first review previously published reports that describe the identification of kaempferol in textiles of the cultural heritage. Then, a case study is presented where the identification of kaempferol in samples removed from three religious textiles, preserved in monasteries of Mount Athos, is described. Kaempferol is detected in mixture with several other colouring components using high performance liquid chromatography (HPLC) with diode-array detection (DAD). Finally, a brief overview of the various colourants found in historical textiles from Mount Athos is provided.

[#] This chapter was previously published in *Kaempferol: Chemistry, Natural Occurrences and Health Benefits*, edited by Garion Villers and Yves Fougere, New York: Nova Publishers Inc., 2013.

[*] Corresponding author. E-mail: y.karapanagiotis@aeath.gr.

In: Biochemistry Research Summaries. Volume 1
Editors: James T. Walker and Catherine M. Jones
ISBN: 978-1-53610-895-8

Chapter 72

GALLIC ACID AS A SOURCE TO USE FOR INCREASING FUNCTIONAL PROPERTIES OF FOOD PRODUCTS[#]

Irene Dini*
Dipartimento di Chimica delle Sostanze Naturali, Università di Napoli "Federico II", Naples, Italy

RESEARCH SUMMARY

Gallic acid and its structurally related compounds are found widely distributed in fruits, plants, vegetables, and derivatives. Esters of gallic acid have a diverse range of industrial uses, as antioxidants in food, in cosmetics and in the pharmaceutical industry. It has drawn particular attention since it has been demonstrated to have many medicinally and therapeutically important properties such as antimicrobial, antiviral, and antitumor. Moreover it has the ability to scavenge peroxyl and DPPH radicals and antioxidant activity at stomach pH. In this chapter, studies of the effects of gallic acid and its derivatives on pathogens, enhancement of cell viability and functionality of probiotics, and their potential applications in food safety and human health are investigated.

[#] This chapter was previously published in *Handbook on Gallic Acid: Natural Occurrences, Antioxidant Properties and Health Implications*, edited by Michelle A. Thompson and Parker B. Collins, New York: Nova Publishers Inc., 2013.

* Email: irdini@unina.it.

In: Biochemistry Research Summaries. Volume 1
Editors: James T. Walker and Catherine M. Jones
ISBN: 978-1-53610-895-8

Chapter 73

GALLIC ACID IMPLICATIONS IN HEALTH: MULTI-THERAPEUTIC AND PROTECTIVE AGENT[#]

Antonio Francesko, Margarida M. Fernandes and Tzanko Tzanov [*]
Group of Molecular and Industrial Biotechnology, Department of Chemical Engineering, Universitat Politècnica de Catalunya, Terrassa, Spain

RESEARCH SUMMARY

Gallic acid is a naturally occurring polyphenol widely distributed in vegetables and fruits ubiquitous in diet. This trihydroxybenzoic acid is most frequently encountered in ester form as a basic structural unit of many health beneficial plant polyphenols. Due to the diverse range of biological effects, gallic acid and its derivatives have been extensively studied for implication in human health. Its well-known potent antioxidant and free radical scavenging properties are conventionally exploited for the therapy of diseases related with oxidative stress, whereas plant polyphenols are generally active against wide range of microbial infections. Moreover, recent experimental and epidemiological studies evidenced an even broader therapeutic potential of gallic acid such as for prevention and treatment of neurodegenerative processes or long-term anti-inflammatory effect for protection against different chronic disorders. In addition, gallic acid elicits chemopreventive effect by inhibiting, retarding and reversing the process of carcinogenesis without harming healthy cells. The present chapter summarises the relevance of gallic acid to human health emphasising the most recent advances in its application as a preventive and therapeutic agent for various diseases. New medical care strategies involving this polyphenol are also reviewed.

[#] This chapter was previously published in *Handbook on Gallic Acid: Natural Occurrences, Antioxidant Properties and Health Implications*, edited by Michelle A. Thompson and Parker B. Collins, New York: Nova Publishers Inc., 2013.

[*] Corresponding author: tzanko.tzanov@upc.edu.

In: Biochemistry Research Summaries. Volume 1
Editors: James T. Walker and Catherine M. Jones

ISBN: 978-1-53610-895-8

Chapter 74

GALLIC ACID: THERMAL AND ANTIOXIDANT PROPERTIES[#]

A. M. T. M. Cordeiro[1], S. S. Damasceno[2], J. G. M. Costa[2], R. Rosenhaim[3], L. E. B. Soledade[4], A. G. Souza[5] and N. A. Santos[6,7]

[1]Campus Manaus-Zona Leste, Instituto Federal de Educação Ciência e Tecnologia do Amazonas, Manaus, AM, Brazil
[2]Departamento de Química Biológica, CCBS, Universidade Regional do Cariri, Crato, CE, Brazil
[3]Departamento de Engenharia Química, CT, Universidade Federal da Paraíba, João Pessoa, PB, Brazil
[4]Campus de Pinheiro, Universidade Federal do Maranhão, Pinheiro, MA, Brazil
[5]Departamento de Química, CCEN, Universidade Federal da Paraíba, João Pessoa, PB, Brazil
[6]Departamento de Tecnologia Sucroalcooleira, CTDR, Universidade Federal da Paraíba, João Pessoa, PB, Brazil
[7]Instituto UFPB de Desenvolvimento da Paraíba, Universidade Federal da Paraíba, João Pessoa, PB, Brazil

RESEARCH SUMMARY

Gallic acid belongs to a wide class of naturally occurring phenolic compounds. It is mainly found in the fruits, leaves, twigs, bark and wood of several plants. It presents with a variety of biological functions, displaying anti-inflammatory, cardiovascular and anticarcinogenic properties, in addition to its antioxidant activity. Epidemiological evidence suggests that an antioxidant-rich diet plays an important role in cancer prevention. This

---

[#] This chapter was previously published in *Handbook on Gallic Acid: Natural Occurrences, Antioxidant Properties and Health Implications*, edited by Michelle A. Thompson and Parker B. Collins, New York: Nova Publishers Inc., 2013.

antioxidant activity is considered as paramount in the inhibition of free radical-related diseases. The action of phenolic compounds, such as gallic acid, is carried out through the inhibition and/or delayed production of free radicals generated by cellular metabolism or exogenous sources. Gallic acid action aims at preventing the attacks of these free radicals on lipids, amino acids of the proteins, double bonds of the polyunsaturated fatty acids and/or the base-paring of DNA, thus avoiding the formation of lesions and also the loss of cellular integrity. Moreover, gallic acid displays scavenging properties that inhibit reactions of iron and copper via metal chelation. Gallic acid is also widely used as a starting material in the synthesis of drugs in pharmaceutical industries. Due to the widespread application of gallic acid, it is important to identify its main natural sources, determine the antioxidant potential and obtain specific knowledge concerning gallic acid's structural features and chemical properties. These are important benchmarks for the roles of antioxidant, since they evaluate the resistance to thermal decomposition and antioxidative power, inhibiting the free radical generation through donations of protons and/or metal chelation.

In: Biochemistry Research Summaries. Volume 1
Editors: James T. Walker and Catherine M. Jones
ISBN: 978-1-53610-895-8

Chapter 75

GALLIC ACID: APPLICATIONS, ANALYSIS AND ELECTROCHEMICAL CHARACTERISATION[#]

E. Skeva and S. Girousi*
Analytical Chemistry Laboratory, School of Chemistry,
Aristotle University of Thessaloniki, Thessaloniki, Greece

RESEARCH SUMMARY

Electrochemical DNA biosensors are promising tools for fast, inexpensive and simple in vitro analysis for the determination of free radicals and antioxidants. Reactive Oxygen Species (ROS) and especially hydroxyl radicals (•OH), are known to damage DNA by oxidation of the bases, which results in their destruction, release or attack of the deoxyribose moieties by strand breaking. These lesions are related to the aging process, cancer, Alzheimer's disease and atherosclerosis. In living systems, hydroxyl radicals are generated from the metal (ferrous or cupric ions) ion-dependent breakdown of hydrogen peroxide by Fenton' s reaction, which is an important mediator of oxidative damage in vivo. Antioxidants such as phenolic acids and plant extracts containing high amounts of them, act as free radical terminators and reduce the effect of the oxidative damage on dsDNA.

In the present work, an overview of natural occurrences, antioxidant properties and health implications is being thoroughly considered.

[#] This chapter was previously published in *Handbook on Gallic Acid: Natural Occurrences, Antioxidant Properties and Health Implications*, edited by Michelle A. Thompson and Parker B. Collins, New York: Nova Publishers Inc., 2013.

* Email: girousi@chem.auth.gr.

In: Biochemistry Research Summaries. Volume 1
Editors: James T. Walker and Catherine M. Jones

ISBN: 978-1-53610-895-8

Chapter 76

THE ANTI-OXIDATIVE AND ANTI-INFLAMMATORY ROLES OF GALLIC ACID ON TRANSCRIPTIONAL REGULATION[#]

***Nando Dulal Das*[*]*, Amitabh Das*[*] *and Young Gyu Chai*[†]**
Department of Molecular and Life Sciences, Hanyang University,
Ansan, Korea

RESEARCH SUMMARY

Gallic acid, a polyhydroxy phenolic compound, is found in various natural products such as tea leave, gallnuts, grapes, green tea, strawberries, lemons, pineapples and in red and white wines. Phenolic compounds are known to donate a proton from their hydroxyl (O–H) bond through hemolytic cleavage and form a stable phenoxy radical. Compounds possessing more than one phenolic hydroxyl group are further shown to generate an even more stabilized phenoxyl radical. As a result, gallic acid is regarded as an excellent antioxidant with a high free radical (reactive oxygen species or reactive nitrogen species-origin) scavenging effect. Among other properties, the anti-inflammatory, anti-bacterial, and antitumor properties are crucial. In this chapter, we mainly focus on the anti-oxidative and anti-inflammatory roles of gallic acid on transcriptional regulation of various cellular regulators. We will describe how gallic acid affects the suppression of anti-oxidative and anti-inflammatory regulators through the controlled regulation of key transcription factors. Since gallic acid and its esters can inhibit the TNFα-induced nuclear translocation of nuclear factor kappa-B (NF-κB) by way of a mechanism independent of inhibitors of kappa-B (IκB) degradation, we also deeply correlate the effect of gallic acid on the NF-κB-mediated transcriptional regulation of various

---

[#] This chapter was previously published in *Handbook on Gallic Acid: Natural Occurrences, Antioxidant Properties and Health Implications*, edited by Michelle A. Thompson and Parker B. Collins, New York: Nova Publishers Inc., 2013.

[*] These authors contributed equally to this work.

[†] Corresponding Author, Young Gyu Chai (Ph.D.) Department of Molecular and Life Sciences, Hanyang University, 1271 Sa 3-dong, Ansan, Gyeonggi-do, 426-791, Korea. Tel: 82-31-400-5513, Fax: 82-31-436-8173, Email: ygchai@hanyang.ac.kr.

early and late inflammatory mediators involved in related disease pathologies. In addition, we will focus on gallic acid's role on the epigenetic level since it exhibits potent p300/CBP-mediated anti-histone acetyl transferase (HAT) activity that inhibits the p65 acetylation-dependent NF-κB activation and production of inflammatory markers. Moreover, this review will show various gallic acid-mediated molecular targets, which affect signaling pathways within cells in different ways involving genetic or epigenetic mechanisms, and engage functional proteins in the formation of inhibiting complexes of strong affinities, ultimately perturbing the expression of oxidative and inflammatory molecules.

In: Biochemistry Research Summaries. Volume 1
Editors: James T. Walker and Catherine M. Jones
ISBN: 978-1-53610-895-8

Chapter 77

GALLIC ACID EXTRACTION AND ITS APPLICATION TO PREVENTION AND TREATMENT OF CANCER[#]

Mayela Govea-Salas,[1] Marily González-Castillo,[1] Cristóbal N. Aguilar,[1] Raúl Rodríguez-Herrera,[1] Alejandro Zugasti-Cruz,[1] Sonia Y. Silva-Belmares[1] and Jesús A. Morlett-Chávez[1,2,*]

[1]Department of Food Research, School of Chemistry, Universidad Autónoma de Coahuila, Saltillo, Coahuila, México
[2]Laboratory of Clinical Analysis and Molecular Diagnostics, School of Chemistry, Universidad Autónoma de Coahuila, Saltillo, Coahuila, México

RESEARCH SUMMARY

Gallic Acid (GA) is one of the most important compounds of the hydrolysable polyphenols group; and can be found in a variety of plant tissues. Extraction of this compound has been performed using optimized conditions such as temperature and solvent type used that ensure the best yields without affecting its properties. Polyphenols are associated with color and flavor of fruits and vegetables; however, the nutritional value of this compound is that it may promote a wide variety of biological activities. Also, it has been reported that Gallic acid participates in regulation of damage by inflammatory diseases, microbial infections, cardiovascular conditions as well as antitumor and anticancer protection. These effects occur because it is a potent antioxidant, and can delay, inhibit, or prevent

---

This chapter was previously published in *Handbook on Gallic Acid: Natural Occurrences, Antioxidant Properties and Health Implications*, edited by Michelle A. Thompson and Parker B. Collins, New York: Nova Publishers Inc., 2013.

* Corresponding author: Jesús A. Morlett-Chávez. Department of Food Research, School of Chemistry. Laboratory of Clinical Analysis and Molecular Diagnostics, School of Chemistry, Universidad Autónoma de Coahuila. C.P. 25282, Saltillo, Coahuila, México. Ph. +52 (844) 4-15-95-34, 4-15-57-52. E-mail: antoniomorlett@uadec.edu.mx.

oxidation of oxidizable materials by scavenging free radicals and decreasing oxidative stress. For this reason several investigations are aimed at the evaluation of phenolic compounds like gallic acid for prevention or treatment of degenerative diseases such as cancer. Results have demonstrated that polyphenols affect some mechanisms of tumor developed in various stages, including initiation, promotion and progression of cancer, which triggers the metastasis process. The objective of the present chapter is to present a review about extraction process of gallic acid as well as the steps and factors implicates in this extraction, properties of this phenolic acid and its applications for cancer treatment as an alternative for the control of the one disease of high incidence, worldwide.

In: Biochemistry Research Summaries. Volume 1
Editors: James T. Walker and Catherine M. Jones
ISBN: 978-1-53610-895-8

Chapter 78

Gallic Acid: Occurrence in Plant Foods and Effects of Agricultural Practices, Vegetative Stage and Processing[#]

Antonietta Baiano*
Dipartimento di Scienze Agrarie, degli Alimenti e dell'Ambiente,
University of Foggia, Foggia, Italy

Research Summary

Gallic acid or trihydroxybenzoic acid belongs to the class of phenolic acids, has antioxidant and antimicrobial properties, and shows cytotoxicity against cancer cells. It occurs in different amounts in a great number of fruits and vegetables and, thus, in their food derivatives. Gallic acid can also arise from the breakdown of tannins. Berries supply gallic acid together with anthocyanins. A cup of hot chocolate contains higher amounts of this compound than beverages such as wine and tea. Other sources of gallic acid are walnuts, hops, apples, and flax seeds.

Agricultural practices, vegetative stage and processing parameters affect the phenolic content and profile of foods. Furthermore, processing operations including washing, cutting, size reduction, mechanical separations, and thermal treatments can modify the gallic acid content.

This review represents a survey of the gallic acid content of plants and plant-based foods and highlights the effects of agricultural practices, vegetative stage, and processing on the content of this phenolic acid.

[#] This chapter was previously published in *Handbook on Gallic Acid: Natural Occurrences, Antioxidant Properties and Health Implications*, edited by Michelle A. Thompson and Parker B. Collins, New York: Nova Publishers Inc., 2013.

* a.baiano@unifg.it, Via Napoli, 25 - 71122 Foggia, Italy.

In: Biochemistry Research Summaries. Volume 1
Editors: James T. Walker and Catherine M. Jones
ISBN: 978-1-53610-895-8

Chapter 79

GALLIC ACID: OCCURRENCES, ANTIOXIDANT ACTIVITY AND HEALTH IMPLICATIONS[#]

Miriam Rejane Bonilla Lemos[1,*] and Rui Carlos Zambiazi[2]
[1]Health Sciences Postgraduation Program, Health Sciences Faculty, Campus Universitário Darcy Ribeiro, Universidade de Brasília, Brasília, DF, Brazil
[2]Posgraduate Program of Science and Agro Industrial Technology, Laboratory of Chromatography. Center of Chemical, Farmaceutical and Food Science, Universidade Federal de Pelotas (UFPel), RS, Brazil

RESEARCH SUMMARY

Studies confirm that oxidative stress may be involved in the etiology of several chronic diseases, episodes genetic, metabolic and neurodegenerative processes such as cardiovascular, diabetes, cancer, Alzheimer's, among others, also working in the acceleration of aging and apoptosis. However it is known that consumption of fruit and vegetables have shown a positive correlation reducing the chances of developing pathologies.

The aim of this study is to relate the bioactive compounds contained in foods derived from plants, among these, the almond baru [Dipteryx alata Vog.], a fruit of savannah, the Cerrado and its relationship to antioxidant activity as a prevention of chronic degenerative diseases and their complications. Almonds were obtained in local trade Brasilia-DF, from three regions of the Cerrado (MT, MG and GO), which were previously selected and randomly assigned into two groups: raw almonds and almond raw film without film. Part of the kernels of these two groups was packaged in transparent polyethylene and stored at -80ºC.

[#] This chapter was previously published in *Handbook on Gallic Acid: Natural Occurrences, Antioxidant Properties and Health Implications*, edited by Michelle A. Thompson and Parker B. Collins, New York: Nova Publishers Inc., 2013.

[*] Corresponding author: Health Sciences Postgraduation Program, Health Sciences Faculty; Campus Universitário Darcy Ribeiro, Universidade de Brasília, Brasília, DF, Brazil. P.O. Box 70910900. e-mail: rebonilla@ gmail.com.

The remainder of almonds with and without film was subjected to roasting, which was held in an oven without air circulation at 150°C/45 minutes. The raw almonds and toast were crushed and analyzed in triplicate.

Thus, knowing that the cellular components are not fully protected by endogenous mechanisms is well established that antioxidants obtained from nature, are indispensable for the proper defense against oxidation and thus begin to play an important role in maintaining health, as These, and contain essential nutrients and micronutrients, contribute to the compounds present in the leaves, fruits, films, and seeds, from its chemical structure, promote action varied in the body. Certain phenolic substances such as flavonoids and phenolic acids, are considerably more potent antioxidant than vitamin C and vitamin E. Gallic acid in a group of phenolic compounds simpler and is inserted in the group of hydroxybenzoic acids or non-flavonoids, being commonly in grapes, strawberries, nuts, cereals and leguminous seeds that act as potential antioxidants with proven preventive action degenerative processes in cardiovascular, and activity against some mutagenic chemicals, striking chemopreventive action, antiviral and antibacterial.

Recent studies show that daily consumption of almond baru [Dipteryx alata Vog.], a seed of edible oilseed feature, reduces oxidative stress induced in rats.

It was confirmed these kernels, the presence of substances with a high content of bioactive compounds, wherein, of the eight phenolic compounds identified by liquid chromatography (HPLC), at concentrations ranging between 67.7 and 224.0 mg/100 acid, gallic was the predominant compound and found that a thermal process is not capable of reducing the content of phenol. Furthermore, the antioxidant activity involves a set of intrinsic factors that protects the body against harmful damage, being able to prevent and heal pre-existing diseases.

In: Biochemistry Research Summaries. Volume 1
Editors: James T. Walker and Catherine M. Jones
ISBN: 978-1-53610-895-8

Chapter 80

ANTIOXIDANT, ANTITUMORAL AND ANTI-INFLAMMATORY ACTIVITIES OF GALLIC ACID[#]

Claudriana Locatelli[1], Fabíola Branco Filippin-Monteiro[2], Ariana Centa[2] and Tânia Beatriz Creczinsky-Pasa[2*]
[1]Curso de Farmácia, Universidade do Oeste de Santa Catarina, Videira, SC, Universidade do Alto Vale do Rio do Peixe, Caçador, SC, Brazil
[2]Departamento de Ciências Farmacêuticas, Universidade Federal de Santa Catarina, Florianópolis, SC, Brazil

RESEARCH SUMMARY

Previous studies have demonstrated a variety of biological activity of gallic acid (GA) including antitumoral, antioxidant and anti-inflammatory properties. Among these numerous data, *in vitro* assays showed that the GA and its derivatives are active against several types of tumor cells. Particularly, the studies have shown that the antitumor activity seems to be related to the induction of apoptosis involving different signalling pathways. Apoptosis induced by GA may be associated with oxidative stress derived from reactive oxygen species (ROS) and mitochondrial dysfunction. Interestingly, GA has both pro-oxidant and antioxidant properties. Intracellular ROS levels were increased in cancer cells treated with GA and their derivatives and these results suggested that GA plays a central role as a pro-oxidant in cancer cell models.

It also has been reported that the cell death promoted by GA in different cell lines may be related with glutathione (GSH) depletion. Since the intracellular GSH has a decisive effect on anticancer drug-induced apoptosis, those apoptotic effects were inversely related to GSH

[#] This chapter was previously published in *Handbook on Gallic Acid: Natural Occurrences, Antioxidant Properties and Health Implications*, edited by Michelle A. Thompson and Parker B. Collins, New York: Nova Publishers Inc., 2013.

[*] Corresponding author: Tel: + 55 48 3721 2212; fax: + 55 48 3721 9542. E-mail: tania.pasa@ufsc.br; taniabcp@gmail.com.

content. The reduction of GSH levels might be related to the capability of GA to decrease the expression, as well as the activity of γ-glutamyl cysteine synthase (γ-GCS).

Regarding tumorigenesis, it has been widely reported that GA is responsible for suppression of tumor angiogenesis. The mechanisms underlying this event might be related to alterations in the interaction among the stromal cell-derived factor-1 (SDF-1), an alpha-chemokine, and CXCR4, a G-protein-coupled receptor. These changes are involved in angiogenesis, upregulating key angiogenic factors including vascular endothelial growth factor (VEGF). Furthermore, GA suppressed ADAM17 expression, a transmembrane protease, which may be associated with the inhibition of invasiveness through the inactivation of the well-known cell-survival signalling pathways PI3K/Akt and Ras/MAPK. Additionally, GA might inhibit the activity of matrix metalloproteinases (MMP) involved in tumor angiogenesis and metastasis, such as MMP-2 and MMP-9. These MMP are directly involved in degradation of extracellular matrix and play pivotal roles in cancer cell migration and invasion.

Finally, GA induced caspase-3 and -8 expression, alteration in the Bcl-2/Bax ratio, and an inhibition in tyrosine phosphorylation by BCR/ALB kinase, as well as a down- regulation in COX-2 levels in leukemia cell lines. The GA was also able to inhibit COX-1 and COX-2 activities in these cell lines, suggesting an anti-inflammatory activity of the compound. Increased COX-2 expression seems to be involved in the development of cancer by triggering cell division and inhibiting apoptosis. These studies suggest that GA induces cell death mediating multivariant mechanisms, being a good candidate to clinical studies for cancer treatment.

In: Biochemistry Research Summaries. Volume 1
Editors: James T. Walker and Catherine M. Jones
ISBN: 978-1-53610-895-8

Chapter 81

BIOTECHNOLOGICAL PRODUCTION OF GALLIC ACID[#]

P. Aguilar-Zárate, M. L. Chávez-González, R. Rodríguez-Herrera and C. N. Aguilar*
Department of Food Science and Technology, School of Chemistry,
Universidad Autónoma de Coahuila, Venustiano, Caranza,
Colonia República Oriente, Saltillo, Coahuila, México

RESEARCH SUMMARY

Gallic acid (3, 4, 5-trihydroxybenzoic acid), is pytochemical compound occurring as a free molecule or as part of tannic acid molecule in plant tissue. Gallic acid has many significant biological activities such as antioxidant, anti-inflammatory, antifungal, and carcinogenic properties, it has attracted considerable interest Tannic acid biodegradation is the key condition for the gallic acid production. Biodegradation of tannic acid can be controlled by enzymatic or microbiological ways, giving the possibility for developing biotechnological processes for the production of gallic acid. Several studies on biotechnological production of gallic acid have generated significant achievements in recent years, interest in the search for new routes to produce more "*clean*" compounds has turned the interest of researchers for fermentation processes, biotransformation rich substrates tannins by microorganisms has led to high yields of gallic acid, however, these new areas of opportunity still present challenges which need studies involve more factors in the microbial production of gallic acid is necessary to make efforts to deepen scale up production studies. In addition, methods for higher recovery and purification of this important compound are necessary.

[#] This chapter was previously published in *Handbook on Gallic Acid: Natural Occurrences, Antioxidant Properties and Health Implications*, edited by Michelle A. Thompson and Parker B. Collins, New York: Nova Publishers Inc., 2013.

* Corresponding author: C. N. Aguilar. Departament of Food Science and Technology. School of Chemistry. Universidad Autónoma de Coahuila. Venustiano Caranza. Colonia República Oriente. CP. 25280. Saltillo, Coahuila, México. E-mail: cristobal.aguilar@uadec.edu.mx.

In: Biochemistry Research Summaries. Volume 1
Editors: James T. Walker and Catherine M. Jones
ISBN: 978-1-53610-895-8

Chapter 82

GALLIC ACID AS A USEFUL INDICATOR OF THE ANTIOXIDANT CAPACITY OF LIQUID FOODS?[#]

F. J. Barba, M. J. Esteve and A. Frigola*
Department of Nutrition and Food Chemistry, Universitat de València, Burjassot, Spain

RESEARCH SUMMARY

In terms of public health, drinking fruit and vegetable juices may well be as effective as consuming whole fruits and vegetables with regard to reducing the risk of chronic disorders. Many of their health benefits can be attributed to phenolic compounds. Phenolic compounds are aromatic metabolites of plants secondary metabolism that have a common structure with an aromatic ring with at least one hydroxyl group, which provides the ability to neutralize reactive species, helping the body to protect itself from oxidative stress. Phenols contribute to fruits' color and taste and have been described as possessing anticarcinogenic and antimutagenic activity. Furthermore, they have been shown to be good contributors to the total antioxidant capacity of foods, resulting in a beneficial effect to human health.

Due to the growing popularity of phenolic antioxidant over the past two decades, an increasing interest in determining the antioxidant activities exhibited by phenolic compounds and their derivatives should also be noted. In this line, special attention has been given to gallic acid. Gallic acid is a naturally abundant plant phenolic compound which is present in liquid foods derived from fruits and vegetables and has been found to exhibit antioxidative properties. Therefore, at this stage of development, there is a need to establish the relationship between phenolic compounds, including gallic acid, with the antioxidant capacity and total phenolic compounds of these food products.

This chapter was previously published in *Handbook on Gallic Acid: Natural Occurrences, Antioxidant Properties and Health Implications*, edited by Michelle A. Thompson and Parker B. Collins, New York: Nova Publishers Inc., 2013.
* E-mail: ana.frigola@uv.es. Phone: +34 963544955, Fax: +34 963544954.

In: Biochemistry Research Summaries. Volume 1
Editors: James T. Walker and Catherine M. Jones

ISBN: 978-1-53610-895-8

Chapter 83

ANTIOXIDANT AND CANCER PROTECTIVE EFFECTS OF GALLIC ACID *IN VITRO* AND *IN VIVO*[#]

Franziska Ferk[1], Armen Nersesyan[1], Miroslav Mišík[1], Karl-Heinz Wagner[2] and Siegfried Knasmüller[1*]

[1]Institute of Cancer Research, Department of Medicine I, Medical University of Vienna, Austria

[2]Department of Nutritional Sciences, University of Vienna, Austria

RESEARCH SUMMARY

Recent findings of our group indicate that a small amount of gallic acid (GA 12 mg/P/d) protects humans against DNA damage and induces the activities of antioxidant enzymes. Furthermore, we saw in animal studies that similar doses prevent damage of the genetic material in inner organs of rats and the induction of preneoplastic hepatic foci by γ-radiation. These promising findings stimulated us to evaluate the current knowledge on ROS and cancer protective properties of the phenolic. Approximately a dozen *in vitro* studies have been published which concern the antioxidant properties of GA; with high concentrations sometimes pro-oxidant effects were observed. Also a few *in vivo* results which were obtained with laboratory rodents are available which indicate GA induces ROS protective enzymes possibly *via* induction of the transcription factor Nrf2. In regard to the cancer protective properties of the compound, it is notable that *in vitro* experiments showed that GA inhibits cell cycle progression at G0/G1 and also at G2/M, (via modulation of the transcription of cyclines) and induces apoptosis in various human cancer cell lines (possible modes of action are inhibition of the synthesis of DNA bases, induction of NfκB, depletion of glutathione and release of ROS). Furthermore, was found that the phenolic decreases the motility and adhesion of cancer cells, inhibits metastasis and prevents angiogenesis *via* inhibition of

---

[#] This chapter was previously published in *Handbook on Gallic Acid: Natural Occurrences, Antioxidant Properties and Health Implications*, edited by Michelle A. Thompson and Parker B. Collins, New York: Nova Publishers Inc., 2013.

[*] Corresponding Author: Siegfried Knasmüller. E-Mail: siegfried.knasmueller@meduniwien.ac.at; Tel: +43 1 40160 57561; Fax: +43 1 40160 957500.

VEGF. Also data from animal experiments (xenograft models with mice) show that the compound prevents the tumour growth. Most of these findings were obtained with doses which are substantially higher as the daily intake of GA *via* the diet which is in Middle Europe in the rage of 18 mg/P/d; however some findings were seen with lower levels that can be easily reached after increase of the consumption by dietary measures. Taken together the current data indicate that GA is an extremely potent antioxidant in humans and may protect against malignant diseases.

In: Biochemistry Research Summaries. Volume 1
Editors: James T. Walker and Catherine M. Jones

ISBN: 978-1-53610-895-8

Chapter 84

APPLICATION OF SPECTROSCOPIC TECHNIQUES FOR THE STUDY OF GALLIC ACID AUTOXIDATION#

Goran M. Nikolić[1], Aleksandar M. Veselinović[1] and Ružica S. Nikolić[2]

[1]Department of Chemistry, Faculty of Medicine, University of Niš, Serbia
[2]Department of Chemistry, Faculty of Natural Sciences and Mathematics, University of Niš, Serbia

RESEARCH SUMMARY

Gallic acid is a polyphenolic compound widely distributed in plants ether in the free form or as a part of more complex polyphenolic compounds. It has a range of biological activities but is best known for its antioxidative activity. All the biological effects of gallic acids in living cells may be connected to the ease of its oxidation or/and autoxidation. Various methods may be used for the study of such processes but spectroscopic methods seem to be especially well suited for such purpose. Out of many spectroscopic methods used for the study of oxidation/autoxidation of various phenolic compounds UV/Vis spectrophotometry and electron spin resonance (ESR) have some particular advantages. UV/Vis spectrophotometry enables direct study of the oxidation/autoxidation of polyphenolic compounds but it lacks the specificity of other spectroscopic techniques. This disadvantage may be overcomed by the application of various chemometric techniques for spectra processing. UV/Vis spectrophotometry is also one of the most economic spectroscopic techniques found in virtually all analytical laboratories and is readily used for detection in liquid chromatography, usually as a HPLC-DAD (high performance liquid chromatography with diode array detection). ESR spectroscopy allows direct detection and characterization of transient free radical species formed in the course of polyphenolic compounds transformations under various conditions. In this chapter we present the results of the

---

This chapter was previously published in *Handbook on Gallic Acid: Natural Occurrences, Antioxidant Properties and Health Implications*, edited by Michelle A. Thompson and Parker B. Collins, New York: Nova Publishers Inc., 2013.

spectroscopic studies of the gallic acid autoxidation under various conditions with special emphasis on the results obtained by the UV/Vis spectrophotometry with chemometric spectral processing and ESR spectroscopy and their application for a better understanding of mechanisms of gallic acid transformation and its action upon living cells.

In: Biochemistry Research Summaries. Volume 1
Editors: James T. Walker and Catherine M. Jones
ISBN: 978-1-53610-895-8

Chapter 85

GALLIC ACID BIOAVAILABILITY IN HUMANS[#]

Andriana C. Kaliora, Panagiotis T. Kanellos and Nick Kalogeropoulos*

Laboratory of Chemistry-Biochemistry-Physical Chemistry of Foods, Department of Science of Nutrition–Dietetics, Harokopio University, Athens, Greece

RESEARCH SUMMARY

Polyphenols are extensively metabolized either in tissues, once they are absorbed through the gut barrier or, for the non-absorbed fraction and the fraction re-excreted in the bile, by the colonic microflora. All polyphenols are conjugated in our body to form O-glucuronides, sulphate esters and O-methyl ethers. The formation of anionic derivatives by conjugation with glucuronides and sulphate groups facilitates their urinary and biliary excretion and explains their rapid elimination.

Animal studies first revealed the possible metabolic fate of gallic acid. In rats, rabbits and chickens the major urinary metabolite is 4-O-methylgallic acid, followed by pyrogallol (conjugated and unconjugated). Small amounts of conjugated 2-O-methylpyrogallol were also detected in rats. In sheep, resorcinol glucuronide is the major product of gallic acid metabolism and unconjugated pyrogallol and resorcinol comprise minor urinary metabolites.

Moreover, recent *in vitro* studies have reported phenolic acids to be main metabolites of anthocyanins after fecal fermentation. However, fragmentation of anthocyanins to phenolic acids in humans has not been studied in detail. One recently identified anthocyanin metabolite is 3-O-methylgallic acid, which is presumably the metabolite of petunidin-3-glucoside, but possibly also a demethylation product of malvidin-3-glucoside.

[#] This chapter was previously published in *Handbook on Gallic Acid: Natural Occurrences, Antioxidant Properties and Health Implications*, edited by Michelle A. Thompson and Parker B. Collins, New York: Nova Publishers Inc., 2013.

* Corresponding author: Nick Kalogeropoulos. Tel.: +30 2109549251; fax: +30 2109577050; E-mail address: nickal@hua.gr.

The few existing studies addressing the bioavailability of gallic acid in humans revealed that, compared to other polyphenols, this phenolic compound is extremely well absorbed. The most common metabolite of gallic acid identified in human biological fluids is 4-O-methylgallic acid, which was found to increase significantly in human plasma within the first 4 h after consumption of a single dose of red wine or dealcoholized red wine. Plasma concentrations of free and glucuronidated forms of gallic acid and 4-O-methylgallic acid also increased after ingestion of 50 mg pure gallic acid, while gallic acid was rapidly absorbed after human administration with tea or acidum gallicum. Concerning the excretion of gallic acid, many studies have identified gallic acid and its metabolites in urinary samples. For example, after acute consumption of 3 cups of black tea, gallic acid was not detected in urine of subjects. However, three gallic acid methyl ethers, 4-O-methylgallic acid, 3-O-methylgallic acid and 3,4-O-dimethylgallic acid, were identified. A recent study on the bioavailability of phenolic compounds present in Corinthian raisins (*Vitis vinifera*) –containing 115.7 ng gallic acid per g- indicated that gallic acid is bioavailable in humans, as its concentrations in plasma increased one hour after raisins consumption. Overall, gallic acid concentrations in plasma were found to range between 2.2±0.9 to 9.9±4.6 (SEM) μg/L.

In: Biochemistry Research Summaries. Volume 1
Editors: James T. Walker and Catherine M. Jones
ISBN: 978-1-53610-895-8

Chapter 86

Gallic Acid and Its Derivatives: Occurrence and Identification in High Altitude Edible and Medicinal Plants[#]

J. Mauricio Peñarrieta,[1,*] Patricia Mollinedo,[1] Enzo Aliaga-Rossel,[1,2] José L. Vila[1] and José A. Bravo[1]

[1]Instituto de Investigaciones en Productos Naturales, Carrera de Ciencias Químicas, Universidad Mayor de San Andrés, La Paz, Bolivia
[2]Instituto de Ecología, Universidad Mayor de San Andrés, La Paz, Bolivia

Research Summary

Phenolic compounds are molecules that have one or more hydroxyl groups bound to an aromatic ring. Together with vitamins, phenolic compounds are considered important dietary antioxidants. Thousands of phenolic compounds are found in plants, and they are classified into different types of functional groups. Phenolic compounds play a number of metabolic roles in plants, in growth and reproduction, and in protection against pathogens and external stress, UV radiation and predators for instance.

Gallic acid is trihydroxybenzoic acid, a phenolic acid, also known as 3,4,5-trihydroxybenzoic acid. The presence of this phenolic compound and its derivatives is extensive in nature and it can be found in many plants and foods like fruits, vegetables and cereals.

Bolivia is currently considered among few countries with a high bio-diversity in the world. This diversity is evident from many species of food plants found in so contrasting geographical regions like the Andean mountains or the Amazon flood plain. In particular, the

[#] This chapter was previously published in *Handbook on Gallic Acid: Natural Occurrences, Antioxidant Properties and Health Implications*, edited by Michelle A. Thompson and Parker B. Collins, New York: Nova Publishers Inc., 2013.

* E-mail. mauricio.penarrieta@gmail.com.

Andean region has a variety of endemic and adapted foods like roots, fruits, cereals and pseudo-cereals all growing at high altitude (2000 to 4200 m.a.s.l.). The presence of ancient native populations such as the Aymara, Quechua and Guarani ethnic groups contributed through the centuries to present with their own and particular *savoir-faire* to the adaptation and use of these plants.

As a sample, this part of the world has given to the world's kitchen two of the most celebrated foods in mankind's history -potato and corn- in their many varieties. Endemic Bolivian non-conventional crops have increased dramatically their importance during recent years due to the introduction of Andean foods onto the international market, never as widely open as nowadays. A typical example is quinoa -*Chenopodium quinoa*-, a protein-rich Andean grain currently available and a very common product in European markets.

Bolivia offers a unique environment for researches on high solar UV radiation levels and related phenolics' contents in plants. The altiplano region particularly, is a *sui generis* site (because of its high altitude), where populations and food plants grow in an abundant variety and quantity with a non-negligible percentage in the per-year national statistics.

This chapter presents the results of gallic acid and its derivatives found in food and high altitude plants. These results include an EIMS and NMR spectroscopic approach for gallic acid.

In: Biochemistry Research Summaries. Volume 1
Editors: James T. Walker and Catherine M. Jones
ISBN: 978-1-53610-895-8

Chapter 87

AN OVERVIEW OF THE TECHNIQUES USED TO ESTIMATE THE ANTIOXIDANT ACTIVITY OF GALLIC ACID[#]

Hongyan Li and Ze-Yuan Deng*
State Key Laboratory of Food Science and Technology, Nanchang University, Nanchang, Jiangxi, China
Institute for Advanced Study, Nanchang University, Nanchang, Jiangxi, China

RESEARCH SUMMARY

The oxidative stress on human health has been a serious issue. It may involve ageing, inflammation, metabolic disorders and cancer. Gallic acid is widely distributed in a number of fruits and plants. It possesses antioxidant activity, thus could decrease the oxidative stress in human. It could also possess anti-cancer activity and other bioactivities related to human health.

The techniques used to determine the antioxidant activity of gallic acid included *in vitro* and *in vivo* methods. The available assays are based on different mechanisms (hydrogen atom transfer and electron transfer), bioavailability, metabolism and other factors, which can lead to varied results. Therefore, this chapter is intended to provide a brief update of recent literature on the *in vitro* and *in vivo* methods available for evaluating the antioxidant activities of gallic acid and the antioxidant values of pure compound or food/plants were also compared and discussed.

The chemical-based methods such as total radical trapping antioxidant potential (TRAP) assay, oxygen radical absorbance capacity (ORAC) assay, luminol-photochemiluminescence (PCL) assay, ferric reducing antioxidant power (FRAP) assay, Trolox equivalent antioxidant

---

[#] This chapter was previously published in *Handbook on Gallic Acid: Natural Occurrences, Antioxidant Properties and Health Implications*, edited by Michelle A. Thompson and Parker B. Collins, New York: Nova Publishers Inc., 2013.

* Corresponding author. Tel./fax: +86 791 88304402. E-mail address: zeyuandeng@ hotmail. com.

capacity (TEAC) assay and 1,1-diphenyl-2-picrylhydrazine (DPPH) radical scavenging assay are used to evaluate the antioxidant activities of phytochemicals in food. The cell-based models include the inhibition on the growth of cancer cells, the protective effect of oxidative stress in various cells exogenously exposed to external stimuli and the cellular antioxidant assay (CAA). The high antioxidant activity *in vitro* does not always lead to high *ex vivo* or *in vivo* antioxidant property. Some *in vivo* studies have also shown promising health effects of gallic acid and the food/plants rich in gallic acid. The animal and human experiments are parts of the methods to evaluate *in vivo* antioxidant potentials. The oxidative stress biomarkers in serum or tissue such as glutathione peroxidase (GSH-Px), glutathione (GSH), superoxide dismutase (SOD), catalase (CAT) and lipid peroxides (LPO) could be used to express the *in vivo* antioxidant activity.

In: Biochemistry Research Summaries. Volume 1
Editors: James T. Walker and Catherine M. Jones
ISBN: 978-1-53610-895-8

Chapter 88

MICROBIAL TRANSFORMATION OF LACTOSE: POTENTIAL OF β-GALACTOSIDASES FOR PROBIOTIC AND PREBIOTIC PURPOSES[#]

Gabriela Zárate[1,2*], Gabriel Saez[2] and Adriana Pérez Chaia[1]
[1]Centro de Referencia para Lactobacilos (CERELA-CONICET), San Miguel de Tucumán, Argentina
[2]Universidad de San Pablo-Tucumán (USP-T), Argentina

RESEARCH SUMMARY

In the last decades, the interest of food industry has focused in the development of functional foods because of the high demand of healthy foods by consumers around the world. In this sense, foods containing probiotic microorganisms have received particular attention. One of the firsts beneficial properties ascribed to probiotics is their ability to contribute to the intestinal metabolism of lactose from the diet and the amelioration of intolerance symptoms. This effect is mediated by the enzyme β-D-galactosidase of the microorganisms included in the product that can hydrolyze the disaccharide into the food before consumption, as well as to contribute to the intraintestinal digestion of lactose.

However, β-galactosidases have more biotechnological potentialities with a wider spectrum of application. This enzyme possess, depending on the reaction conditions (substrate concentration, aw, pH, reaction temperature), hydrolase or transferase activities, owing the last one to the production of oligosaccharides containing 2 to 10 monomers obtained by transgalacto-sylation reactions that transfer galactose hydroxyl groups to the disaccharide lactose and/or the different oligomers that are being generated.

[#] This chapter was previously published in *Lactose: Structure, Food Industry Applications and Role in Disorders*, edited by David Green and Emma Lee, New York: Nova Publishers Inc., 2013.

[*] E-mail: gzarate@cerela.org.ar.

Galactooligosaccharides (GOS) are well known prebiotics that modulate the colonic microbiota in a healthy manner and also have other promising industrial applications such as their addition to foods as no-cariogenic, low-calories sweeteners. The raw material used for GOS production could be the lactose from milk as well as the whey discarded by dairy industries from cheeses manufactures. Since whey is considered an environmental contaminant its utilization in this regard seems to be an attractive option.

In the present chapter, the potential of well-known probiotics (such as lactic acid bacteria, bifidobacteria, propionibacteria and yeasts) to contribute to the intestinal metabolism of lactose and/or to produce prebiotic GOS by their β-galactosidase activities is reviewed. In many cases the main biochemical properties of the enzyme of the microorganisms with the higher hydrolytic activities have been determined as well as their abilities to improve lactose digestion in animal models and clinical trials. In this regard, multiple therapeutic strategies have been assessed with different results.

On the other side, the production of GOS by transgalactosylation using *β*-galactosidases from GRAS (Generally Recognized As Safe) microorganisms has also been studied. Depending on the enzyme source and the reaction conditions, the extent of oligosaccharide formation could be as high as 40% of the total sugar content in the solution and three to eleven types of oligosaccharides could be found, including disaccharides other than lactose, trisaccharides and various tetrasaccharides. Although some suitable sources of β-galactosidases that can be used to synthesize novel GOS have been characterized, further studies on the prebiotic effects of these GOS in pure cultures of bifidobacteria, and colonic microbiota of animals and human volunteers are still missing.

Besides, the selection of probiotic microorganisms with appropriate β-galactosidase activities should be complemented with in vivo assays (survival and activity in the intestinal conditions) and certain technological parameters in order to be included in the formulation of novel functional foods, particularly fermented products enriched in GOS and β-galactosidase.

In: Biochemistry Research Summaries. Volume 1
Editors: James T. Walker and Catherine M. Jones

ISBN: 978-1-53610-895-8

Chapter 89

MILK, LACTOSE, LACTASE: THE MEDICAL ADVENTURE[#]

Andrew Szilagyi*
Division of Gastroenterology, Department of Medicine
Jewish General Hospital, McGill University School of Medicine,
Montreal, Quebec, Canada

RESEARCH SUMMARY

Lactose is a disaccharide which is the main carbohydrate present in milk of most mammals, including humans. At birth mammals possess lactase for digestion but lose this enzyme after weaning. In humans however, about 1/4 to 1/3 of the world's population retain lactase throughout adulthood, allowing continued consumption of milk.

The consequences of this human divide are the incurring of digestive symptoms in those who have limited intestinal lactase and consume large enough quantities at a time.

The medical issues arising from this dichotomy then include the consequences of lactose associated symptoms and the more important consequences of diary food avoidance. In addition dairy foods are linked or hypothesized to increase or decrease risks for many diseases. Perhaps one of the most interesting aspects of the world lactase dichotomy is the relationship of population distribution and geographic patterns of modern day, so called "western" lifestyle linked diseases. The asymmetric distribution and increasing incidence/prevalence of these diseases in developing nations sheds light on the possible co evolutionary impact of lactase pheno or genotype on such diseases. The development of the asymmetric lactase patterns and consequent medical aspects will be discussed.

[#] This chapter was previously published in *Lactose: Structure, Food Industry Applications and Role in Disorders*, edited by David Green and Emma Lee, New York: Nova Publishers Inc., 2013.

* Address 3755 Cote St Catherine Rd, Rm E110 H3T 1E2 Montreal Que. Tel 514 340 8144 ; Fax 514 340 8282 ; E-mail : aszilagy@jgh.mcgill.ca.

In: Biochemistry Research Summaries. Volume 1
Editors: James T. Walker and Catherine M. Jones

ISBN: 978-1-53610-895-8

Chapter 90

WHEY LACTOSE FERMENTATION BY KEFIR IMMOBILIZED ON AGRO-INDUSTRIAL WASTES#

Magdalini Soupioni
Nuclear Chemistry Laboratory, Department of Chemistry,
University of Patras, Patras, Greece

RESEARCH SUMMARY

Whey is the main liquid by-product of the dairy manufactures, in which a large quantity of the milk lactose is removed. So, whey consists of lactose (4.8-5.3%), proteins (0.8-1.6%), fats (0.3-1.3%), vitamins (0.8%) and minerals. The fat and protein components are rather easily recovered, but lactose responsible for 90% of whey's COD remains unused.

The last two decades, the natural mixed culture kefir was chosen for whey lactose exploitation, because its microbiological composition ensures simultaneous alcoholic and lactic acid whey fermentations resulting to useful products formation like ethanol, lactic and propionic acid. Also, the immobilization of kefir cells on various natural food grade supports and the preparation of biocatalysts suitable for easy and fast whey fermentation have received considerable attention by researchers.

In this chapter the whey lactose fermentation kinetics by kefir immobilized on Delignified Cellulosic Materials (DCM), Gluten Pellets (GP), Brewer's Spent Grains (BSG), Malt Spent Rootlets (MSR) and Corn Starch Gel (CSG) was examined in relation to free kefir cells (FC) for comparative reasons. Also, lactose uptake rate by immobilized and free kefir cells during fermentation was monitored by adding ^{14}C-labelled lactose in whey and the produced ethanol was determined.

The biocatalysts prepared by immobilization of kefir cells on agro-industrial wastes were proved effective for whey fermentation at 30°C and pH 5.5, reducing significantly fermentation time and increasing lactose uptake rate in comparison to free cells. Specifically, using cheap, abundant and food grade BSG or DCM biocatalysts for whey fermentation is

---

This chapter was previously published in *Lactose: Structure, Food Industry Applications and Role in Disorders*, edited by David Green and Emma Lee, New York: Nova Publishers Inc., 2013.

potentially acceptable by industries as the fermentation time was estimated only about 7 hours in that case. The highest alcohol concentration was observed by MSR biocatalyst.

In: Biochemistry Research Summaries. Volume 1
Editors: James T. Walker and Catherine M. Jones
ISBN: 978-1-53610-895-8

Chapter 91

CRYSTALLINE AND AMORPHOUS LACTOSES: TABLETING PROPERTIES AND THEIR APPLICATION FOR THE PRODUCTION OF PHARMACEUTICAL COMPACTS#

John Rojas, Santiago Hernandez and Alejandro Giraldo
Department of Pharmacy, School of Pharmaceutical Chemistry, University of Antioquia

RESEARCH SUMMARY

Lactose is the most widely used excipient for the manufacture of tablets and capsules due to its relatively low cost, high solubility, good absorption and rapid metabolism in the human body. It is blended with a drug and other excipients and compressed to make tablets for oral administration. However, lactose is not desirable to formulate alkaline drugs or active ingredients with amine moieties due to the formation of a Maillard reaction. It is available as α-lactose monohydrate, anhydrous α-lactose, anhydrous β-lactose and spray-dried lactose.

Spray-dried lactose had the best flow. On the contrary, β-lactose exhibited the poorest flow, but along with the anhydrous α-lactose presented the highest binding ability. Conversely, α-lactose monohydrate presented the poorest compactibility, the fastest disintegration time and its compactibility was not affected by moisture.

Spray-dried lactose is composed of α-lactose monohydrate crystals glued together by amorphous lactose. On the contrary, α-lactose and α/β-lactose were the most fragmenting materials. In turn, these materials were the least sensitive to magnesium stearate. Lactose tablets independent of the crystalline form showed a moisture-dependent tensile strength resulting into a decreased compactibility. However, under storage compacts showed a very small elastic recovery.

This chapter was previously published in *Lactose: Structure, Food Industry Applications and Role in Disorders*, edited by David Green and Emma Lee, New York: Nova Publishers Inc., 2013.

It is concluded that compactibility and compressibility of lactose was related to its polymorphic form.

In: Biochemistry Research Summaries. Volume 1
Editors: James T. Walker and Catherine M. Jones
ISBN: 978-1-53610-895-8

Chapter 92

LACTOSE AND ITS INFLUENCE IN DIFFUSION OF DIVALENT METAL CHLORIDES[#]

Ana C. F. Ribeiro[1,*], *Luis M. P. Verissimo*[1], *Victor M. M. Lobo*[1] *and Miguel A. Esteso*[2,†]
[1]Department of Chemistry, University of Coimbra, Coimbra, Portugal
[2]Departamento de Química I, U.D. Química Física, Facultad de Farmacia, Universidad de Alcalá, Alcalá de Henares, Madrid, Spain

RESEARCH SUMMARY

Lactose (disaccharide derived from the condensation of galactose and glucose) is not only a technological important compound, but it also enjoys biological, medical, pharmaceutical, food and biomedical applications [1-4]. As a typical non-electrolyte carrying hydrophilic hydroxyl groups capable of hydrogen-bonding, its properties play a significant role in the reaction conditions of many current industrial processes, such as enzymatic conversion of biomass to useful chemicals, and in many biochemical interactions with metal ions.

However, while numerous studies have been carried out on the thermodynamic properties of binary aqueous lactose solutions, data are more limited on the transport behaviour of these sugar systems in aqueous solutions. The characterization of the transport properties, particularly diffusion coefficients, is important for fundamental reasons, helping us to understand the nature of the structure of the aqueous systems containing metal ions, and for practical application in fields such as corrosion. In fact, they provide a direct measure of molecular mobility, an important factor in the preservation of biological materials in sugar matrixes.

[#] This chapter was previously published in *Lactose: Structure, Food Industry Applications and Role in Disorders*, edited by David Green and Emma Lee, New York: Nova Publishers Inc., 2013.

[*] Corresponding author: Tel: +351-239-854460; Fax: +351-239-827703; Email: anacfrib@ ci.uc.pt luisve@gmail.com.

[†] Email: miguel.esteso@uah.es.

Lactose is the most important carbohydrate present in the milk of most mammals. As a consequence, for the lactose-intolerant persons it has been verified that the ingestion of enough calcium may be difficult. For these people, who reveal inability or insufficient ability to digest lactose, because of a deficiency of the enzyme lactase, it is often recommended that they have calcium supplements, such as calcium carbonate or others.

Hopefully, the studies reported here will lead to an increase in the know-how, which will allow a better understanding of the physicochemical conditions underlining the diffusion phenomena occurring in different biological systems. Thus, we proposed to extend our studies concerning the experimental determination of the diffusion coefficients for systems containing lactose [5-9], that is, for aqueous system of calcium chloride plus lactose, having in mind a better understanding of the structure of these systems, and of the thermodynamic and transport behaviour of this carbohydrate in those aqueous media. On the other words, the main scope of our work shown here is to show the effect of lactose on the diffusion of metal ions involved in biological systems.

In: Biochemistry Research Summaries. Volume 1
Editors: James T. Walker and Catherine M. Jones
ISBN: 978-1-53610-895-8

Chapter 93

LIGNINOLYTIC PEROXIDASES: UNIQUE PROPERTIES AND BIOTECHNOLOGICAL POTENTIAL[#]

N. Pozdnyakova[1,*], K. Wlizlo[2], K. Szalapata[2], A. Jarosz-Wilkolazka[2] and O. Turkovskaya[1]
[1]Institute of Biochemistry and Physiology of Plants and Microorganisms, Russian Academy of Sciences, Saratov, Russia
[2]Department of Biochemistry, Maria Curie-Sklodowska University, Lublin, Poland

RESEARCH SUMMARY

Ligninolytic fungi (white rot and litter-decomposing fungi) are a physiological rather than taxonomic group connecting fungal strains which are capable of extensively degrading lignin. The unique physiological and biochemical properties permit these fungi to transform and mineralize of wide range of organopollutants with structural similarities to lignin. They produce a complex of extracellular oxidative enzymes, including lignin peroxidase, manganese peroxidase, versatile peroxidase, and laccase, which are involved in degradative processes. This enzyme complex is produced by all ligninolytic fungi but its pattern is dependent on fungal species and cultivation conditions. Lignin peroxidase catalyzes oxidative cleavage of C-C bonds and ether (C-O-C) bonds in non-phenolic aromatic substrates with high redox potential. Manganese peroxidase oxidizes Mn^{2+} to Mn^{3+}, which facilitates the degradation of phenolic compounds or, in turn, oxidizes a second mediator for the breakdown of non-phenolic compounds. Versatile peroxidase is hybrid of lignin peroxidase and manganese peroxidase combining the catalytic possibilities of those two peroxidases. This review presents the most important molecular and catalytic properties of ligninolytic peroxidases including molecular structure, substrate specificity, optimal pH values and

---

[#] This chapter was previously published in *Peroxidases: Biochemical Characteristics, Functions and Potential Applications,* edited by Lucas Bogaert and Noah Coppens, New York: Nova Publishers Inc., 2013.

[*] Corresponding author: nataliapozdnyakova@yahoo.com.

temperature, influence of effectors, catalytic activity towards lignin and xenobiotics. The features of catalytic mechanisms of these enzymes are comparing. The possible functions of ligninolytic peroxidases and supposed dependence the fungal degrading properties on pattern and properties to produce various peroxidases are discussed. Finally the potential of biotechnological use of these enzymes is under consideration.

In: Biochemistry Research Summaries. Volume 1
Editors: James T. Walker and Catherine M. Jones

ISBN: 978-1-53610-895-8

Chapter 94

PLANT PEROXIDASES: FUNCTIONS AND REGULATION DURING PATHOGENESIS[#]

A. V. Sorokan' and I. V. Maksimov
Institute of Biochemistry and Genetics, UFA Scientific Centre,
Russian Academy of Science, Russia

RESEARCH SUMMARY

Peroxidase (donor: H_2O_2oxidoreductase, EC 1.11.1.7) (PO) belongs to a class of widespread and vital enzymes. One of its specific features is the multiplicity of its molecular isoforms exhibiting various functions], and the highly variable genes encoding them are controlled by various *cis*-elements and *trans*-factors that are responsible for their differentiated expression activity depending on environmental conditions and stages of ontogenesis. POs are localizing in cytoplasm, plasma membrane or secrete outside of cell. In cell wall POs presence in the ionic-bound or covalent bound fraction of proteins, and even freely circulated in apoplast. POs are key enzymes of lignification capable of binding with polysaccharides of plant and (or) pathogens. An important physiological sense of these reactions is the strengthening of the cell wall and construction on the pathogens' paths of physical barriers consisting of lignin- and suberin-containing polymers, which also possess high toxicity. This defense reaction is developing more or less automatically, thus these barriers are formed just in the zone of pathogen penetration and concurrently with the beginning of the active pathogen expansion into the tissues of the host plant. However, mechanisms of development of these events are still unclear. Expression of some genes encoding different isoPOs are regulated by pathogens (and their metabolites), elicitors, and stress-related hormones and hormone-like compounds, such as salicylic and jasmonic acids, ethylene and others.

[#] This chapter was previously published in *Peroxidases: Biochemical Characteristics, Functions and Potential Applications,* edited by Lucas Bogaert and Noah Coppens, New York: Nova Publishers Inc., 2013.

In: Biochemistry Research Summaries. Volume 1
Editors: James T. Walker and Catherine M. Jones
ISBN: 978-1-53610-895-8

Chapter 95

PHOSPHOLIPIDS AND PEROXIDASES IN MITOCHONDRIAL APOPTOSIS[#]

Ekaterina A. Korobkova*
Department of Sciences, John Jay College of Criminal Justice, (CUNY),
New York, NY, US

RESEARCH SUMMARY

Phospholipids constitute an essential part of cellular membranes, in particular, inner and outer membranes of mitochondria. They influence the function of mitochondrial proteins and cytosolic proteins associates with mitochondria. Phospholipids regulate the assembly and maintenance of the proteins embedded in the membranes, they modulate enzymes' ox-red properties, electron-transfer properties, kinase activity, and peroxidase activity. Phospholipids thus play a crucial role in all cellular processes such as protein biogenesis, protein import into mitochondria, energy production, and apoptosis. Over the last decade, a significant amount of research efforts has been devoted to mitochondrial apoptosis, in particular, early stages of apoptosis that precede mitochondrial permeability transition pore (PTP) opening and the release of proapoptotic factors into the cytoplasm, specifically cytochrome c (cyt c). Cyt c is a soluble protein located in the intermembrane space of mitochondria whose primary function is the electron shuttle between complexes III and IV of the respiratory chain. Cyt c is attached to the inner mitochondrial membrane (IMM) through a phospholipid molecule, cardiolipin (CL). It was demonstrated that cyt c, whose peroxidase function is negligible in a free state, exhibits a significant peroxidase activity upon binding to CL. This cyt c-CL complex catalyzes the peroxidation of CL. Peroxidized CL stimulates the formation of the permeability pore in the membranes and the release of cyt c into the cytoplasm through these pores. The increase of the peroxidase activity of cyt c in the CL-bound state is thus considered to be a prerequisite for apoptosis. Cells have also developed anti-apoptotic mechanisms. One of these

---

[#] This chapter was previously published in *Peroxidases: Biochemical Characteristics, Functions and Potential Applications*, edited by Lucas Bogaert and Noah Coppens, New York: Nova Publishers Inc., 2013.
[*] Corresponding author: ekorobkova@jjay.cuny.edu.

anti-apoptotic functions is executed by a mitochondrial phospholipid hydroperoxide glutathione peroxidase or GPx4, a protein that defends cells from oxidative injury, attenuates cardiac disfunction, and participates in sperm development. GPx4 plays an essential role in preventing cyt c release from mitochondria. Specifically, it reduces oxidized phospholipids precluding cyt c detachment from CL and consequently suppressing apoptosis. We will describe the structure of cyt c and GPx4 as well as the catalytic cycles of the two peroxidases. We will highlight the recent advances and unresolved issues regarding biochemical mechanisms underlying cyt c-CL interactions. We will discuss the structural characteristics of cyt c-CL interactions responsible for the tight binding, unfolding of cyt c, and the enhancement of its peroxidase activity. We will also discuss how peroxidized CL triggers PTP opening and how the balance between the two peroxidases, CL-bound cyt c and GPx4, affects the early stages of mitochondrial apoptosis.

In: Biochemistry Research Summaries. Volume 1
Editors: James T. Walker and Catherine M. Jones
ISBN: 978-1-53610-895-8

Chapter 96

GLUTATHIONE PEROXIDASES IN STEM, CANCER, AND CANCER STEM CELLS#

Shanchun Guo[1]*, Lily Yang[2], Guangdi Wang[3] and Alexzander A. A. Asea[1]
[1]Department of Microbiology, Biochemistry & Immunology, Morehouse School of Medicine, Atlanta, GA
[2]Department of Surgery, Emory University School of Medicine, Atlanta, GA
[3]Department of Chemistry, Xavier University of Louisiana, New Orleans, LA

RESEARCH SUMMARY

Oxidative stress, as a disturbance in the equilibrium between free radicals (FR), reactive oxygen species (ROS), and endogenous antioxidant defense mechanisms is caused by an imbalance between the production of reactive oxygen and a biological system's ability to readily detoxify the reactive intermediates or easily repair the resulting damage. Mounting evidence has implicated that oxidative stress is involved in various physiological and pathological processes, including DNA damage, proliferation, cell adhesion, and survival of cancer cells. Glutathione peroxidases (GPxs) (EC 1.11.1.9), which can reduce lipid hydroperoxides to their corresponding alcohols and free hydrogen peroxide to water, are enzyme family with peroxidase activity whose main biological roles are to protect the organism from oxidative damage. So far 8 sub-members of GPxs have been identified in humans, i.e. GPx 1–8. They all can react with H_2O_2 and soluble fatty acid hydroperoxides. A large number of publications have demonstrated that GPxs have significant roles in different stages of carcinogenesis. In this book chapter, we will review recent study progress of GPxs in stem, cancer and cancer stem cells. This knowledge will shed light on mechanisms that lead to better modes of cancer treatment.

This chapter was previously published in *Peroxidases: Biochemical Characteristics, Functions and Potential Applications,* edited by Lucas Bogaert and Noah Coppens, New York: Nova Publishers Inc., 2013.

* Corresponding author: Shanchun Guo, E-mail: sguo@msm.edu

In: Biochemistry Research Summaries. Volume 1
Editors: James T. Walker and Catherine M. Jones
ISBN: 978-1-53610-895-8

Chapter 97

BIOCHEMICAL RESPONSES OF AQUATIC MACROPHYTES TO TWO ABIOTIC STRESSES[#]

Takashi Asaeda*, Tanjeena Zaman and Md H. Rashid
Department of Environmental Science, Saitama University,
Sakura, Saitama, Japan

RESEARCH SUMMARY

Exposure to stress can lead to the disruption of cellular, molecular and physiological processes in plants. Plants generate ROS to cope with different forms of abiotic stresses. However, if a plant is exposed to stress conditions for a considerable period, the excess generation of ROS affects its radical scavenging ability in the long run. Thus the effects of environmental variability of a plant are reflected in its growth, reproduction and colonization. Aquatic plants are considered unchangeable biological filters and play an important role in the proper functioning of aquatic ecosystem. They often encounter abiotic stress of varying magnitudes and acquire with different adaptive characters. In this chapter, we will be focusing on the oxidative responses of aquatic macrophytes to various abiotic stresses with a special concern on flow velocity and redox states.

[#] This chapter was previously published in *Peroxidases: Biochemical Characteristics, Functions and Potential Applications,* edited by Lucas Bogaert and Noah Coppens, New York: Nova Publishers Inc., 2013.

* Corresponding author: Dr. Takashi Asaeda, Department of Environmental Science, Saitama University, 255 Shimo-okubo, Sakura, Saitama, 338-8570, Japan, E. mail: asaeda@mail.saitama-u.ac.jp.

In: Biochemistry Research Summaries. Volume 1
Editors: James T. Walker and Catherine M. Jones
ISBN: 978-1-53610-895-8

Chapter 98

PLANT RESPONSES TO STRESSES: ROLE OF ASCORBATE PEROXIDASE IN THE ANTIOXIDANT PROTECTION[#]

I. M. Huseynova[1], D. R. Aliyeva[1] and J. A. Aliyev[1,2]
[1] Department of Fundamental Problems of Biological Productivity, Institute of Botany, Azerbaijan National Academy of Sciences, Baku, Azerbaijan
[2] Department of Plant Physiology and Biotechnology, Research Institute of Crop Husbandry, Ministry of Agriculture of Azerbaijan Republic, Baku, Azerbaijan

RESEARCH SUMMARY

The change in activities and numerous molecular forms of antioxidant enzymes is one of the parameters of plant biological tolerance against constantly changing environmental conditions. A group of genes involved in the synthesis of ascorbate is used in genetic engineering for the improvement of plant tolerance to biotic and abiotic stresses. The level of H_2O_2 in plant cells is controlled by two enzymes: catalase and ascorbate peroxidase (APX), though the key role in decomposition of H_2O_2 belongs to APX. Ascorbate peroxidase converts H_2O_2 into H_2O and O_2. The APX activity and electrophoresis spectra of its isoenzyme content have been investigated in plants subjected to abiotic (soil drought) and biotic (virus infections) stresses. The analysis of electrophoretic spectra of four isoforms in wheat leaves at the beginning of drought (flowering phase) revealed four isoforms. The amounts of APX isoforms increased to 7 during the wax ripeness phase both in irrigated plants and plants subjected to drought. The long-term drought led to the increase in the intensity of high-molecular isoforms and emergence of three additional isoforms of the enzyme. In this instance, the APX activity level remained higher in drought tolerant wheat genotypes

---

[#] This chapter was previously published in *Peroxidases: Biochemical Characteristics, Functions and Potential Applications*, edited by Lucas Bogaert and Noah Coppens, New York: Nova Publishers Inc., 2013.

compared with the sensitive ones. At the same time, in both tolerant and sensitive varieties, enhanced APX activity compared with the control variants was observed at the end of drought, which is apparently related to the synthesis of new molecules of the enzyme (i.e. it is stipulated by the changes of genome expression under stress). The immune enzymatic method revealed the presence of CMV virus in *C. sativus* L. plants and TMV in *L. esculentum* Mill. plants. The obtained results show significant differences in APX activities in all virus infected leaves compared with controls. The effect of the viral infection caused approximately a 1.6-fold increase of the APX activity in *L. esculentum* Mill. and a 2.1-fold increase in *C. sativus* L. compared with corresponding controls. Pathogenesis was accompanied by an increase in the quantity of APX isoforms, which reached 5 in leaves of *L. esculentum* Mill. and by the formation of 2 new isoforms and intensification of the enzyme electrophoresis spectra in infected leaves of *C. sativus* L. compared with the control variants. Thus, an increase in ascorbate peroxidase activity and *de novo* synthesis of its numerous molecular forms under stress conditions testify its involvement in the antioxidant defense system.

In: Biochemistry Research Summaries. Volume 1
Editors: James T. Walker and Catherine M. Jones
ISBN: 978-1-53610-895-8

Chapter 99

ROUTE, DOSE AND DURATION OF EXPOSURE TO CADMIUM-RELEVANCE TO OXIDATIVE STRESS INDUCTION#

Vesna Matović*, Danijela Đukić-Ćosić, Aleksandra Buha and Zorica Bulat
Department of Toxicology "Akademik Danilo Soldatović",
University of Belgrade-Faculty of Pharmacy, Belgrade, Serbia

RESEARCH SUMMARY

Cadmium (Cd) is a toxic metal currently ranked as seventh (out of 275) on ATSDR priority list of hazardous substances and it is considered as one of the most important occupational and environmental pollutants. This carcinogenic metal causes adverse effects in various tissues, particularly kidneys after prolonged exposure. However, the mechanisms of these toxic effects are still not completely understood. Literature data indicate different mechanisms of cadmium toxicity, oxidative stress being one of the most important. Experiments confirm that Cd, metal with no redox potential, can indirectly induce the production of reactive oxygen species (ROS) by affecting antioxidant enzymes (superoxide dismutase, catalase, glutathione peroxidase, glutathione reductase, glutathione S-transferase), nonenzymatic antioxidants (glutathione, total sulfhydryl groups, antioxidative vitamines), and Fenton metals, iron and copper.

This chapter summarizes our investigations on cadmium-induced oxidative stress conducted on three different animal species: rabbits, mice and rats. Application of a single oral dose of 50 mg Cd/kg b.w. in rabbits showed increased levels of lipid peroxidation in blood 4-6 hours after Cd intoxication indicating early development of oxidative stress in conditions of acute Cd poisoning. Experiments carried out on mice showed that single oral

---

This chapter was previously published in *Peroxidases: Biochemical Characteristics, Functions and Potential Applications,* edited by Lucas Bogaert and Noah Coppens, New York: Nova Publishers Inc., 2013.

* Corresponding author Email: vevodi@pharmacy.bg.ac.rs.

exposure to 20 mg Cd/kg b.w. resulted in intensive ROS production in liver while subacute oral dose of 10 mg Cd/kg b.w. induced more pronounced effect in kidneys, thus indicating different effect of Cd intoxication depending on duration of Cd exposure. Further experiments performed on rats provided strong evidence that Cd can induce oxidative stress in both blood and liver after oral (30 mg Cd/kg b.w.) and intraperitoneal (1.5 mg Cd/kg b.w.) acute intoxication, although intraperitoneal administration showed more pronounced negative effects.

Results of these studies demonstrate that oxidative stress is an important mechanism of Cd toxicity and that Cd ability to induce oxidative stress in different tissues depends on the route, dose and duration of exposure to this toxic metal.

In: Biochemistry Research Summaries. Volume 1
Editors: James T. Walker and Catherine M. Jones
ISBN: 978-1-53610-895-8

Chapter 100

HYPOXIA- AND ANGIOGENESIS-RELATED *MYO*-INOSITOL PHOSPHATES: SYNTHESIS AND BIOLOGICAL EVALUATION[#]

Alexandros E. Koumbis[1,*] and Konstantina C. Fylaktakidou[2,†]
[1]Laboratory of Organic Chemistry, Department of Chemistry,
Aristotle University of Thessaloniki, Thessaloniki, Greece
[2]Department of Molecular Biology and Genetics,
Democritus University of Thrace, Alexandroupolis, Greece

RESEARCH SUMMARY

One of the most fundamental physiological processes in the aerobic organisms is the delivery of oxygen to all tissues. Deprivation of oxygen leads to hypoxia is a pathological condition playing a central role in numerous types of diseases, such as cancer (solid tumors), cardiovascular diseases and anemia. In this chapter we review the synthesis of a number of *myo*-inositol phosphates and their biological evaluation regarding hypoxia and angiogenesis in an independent or a combined relationship. Some of the compounds were found to be involved in the increased oxygen delivery by free hemoglobin or whole red blood cells, in the suppression of hypoxia-induced factor 1α (HIF-1α), in the inhibition of the phosphatidylinositol 3-kinase (PI3K) pathway, and in the reduction of angiogenesis. Consequently, the eradication or growth reduction of solid tumors, the prevention of sever anemia caused by sickle cell disease, and the enhanced exercise capacity of subjects with severe heart failure have been demonstrated. These results have evinced a leading role for *myo*-inositol phosphates in the treatment of diseases related to hypoxia and angiogenesis.

[#] This chapter was previously published in *Inositol: Synthesis, Functions and Clinical Implications*, edited by Henrique Rocha and Marina Cardoso, New York: Nova Publishers Inc., 2013.

[*] E-mail: akoumbis@chem.auth.gr.
[†] E-mail: kfylakta@mbg.duth.gr.

In: Biochemistry Research Summaries. Volume 1
Editors: James T. Walker and Catherine M. Jones

ISBN: 978-1-53610-895-8

Chapter 101

INOSITOL DERIVATIVES IN MYCOBACTERIUM TUBERCULOSIS: FUNCTION, BIOSYNTHESIS AND THERAPEUTIC IMPLICATIONS[#]

Xinyi Huang[1] and Marcy Hernick[*,1,2]
[1]Department of Biochemistry, Virginia Tech, Blacksburg, VA
[2]Department of Pharmaceutical Sciences, Appalachian College of Pharmacy, Oakwood, VA

RESEARCH SUMMARY

Inositol (Ins) is a polyol used for the preparation of Ins derivatives that carry out essential functions in eukaryotes. Although rarely found in bacterial species there are Ins derivatives that play critical roles in actinomycetes, a group of Gram-positive bacteria that includes *Mycobacterium* species such as *M. tuberculosis*, the causative agent of tuberculosis (TB). Consequently, there is much interest in understanding the functions and biosynthesis of Ins derivatives in mycobacteria to gain insights that can be used for the development of therapeutic agents. Mycobacteria can either obtain Ins from the environment through Ins transporters or synthesize it *de novo* via a two-step enzymatic pathway. Cellular Ins is converted enzymatically into phosphatidylinositol (PI), a precursor for more complex glycolipids, or incorporated as L-*myo*-inositol 1-phosphate (Ins-1P) into the small molecule mycothiol (MSH). These Ins derivatives carry out critical functions required for mycobacterial viability and virulence. The mycobacterial cell envelop contains an abundance of mannosyl-phosphatidylinositol containing glycolipids and lipoglycans, including phosphatidylinositol mannosides (PIMs) lipoarabinomannan, (LAM), and lipomannan (LM). Importantly, PIMs, LAM, and LM have been shown to function as immunomodulatory molecules, promote the entry of mycobacteria into phagocytic cells, and regulate phagosome maturation. Mycobacteria also produce the Ins derivative MSH, which is used by

---

This chapter was previously published in *Inositol: Synthesis, Functions and Clinical Implications*, edited by Henrique Rocha and Marina Cardoso, New York: Nova Publishers Inc., 2013.

* Corresponding author: email: mhernick@acp.edu.

mycobacteria as the primary reducing agent and in the detoxification of xenobiotics. Due to these important functions, mycobacterial Ins derivatives are of therapeutic interest for vaccine development and as sources of potential targets for drug development. This chapter reviews mycobacterial uptake of Ins and biosynthesis of Ins derivatives, known functions of Ins derivatives in *M. tuberculosis*, and the progress to date on targeting Ins derivatives for vaccine and drug development.

In: Biochemistry Research Summaries. Volume 1
Editors: James T. Walker and Catherine M. Jones

ISBN: 978-1-53610-895-8

Chapter 102

INOSITOLS AND EPILEPSY#

R. Solomonia,[1,2] E. Lepsveridze,[1] T. Kiguradze[2] and M. Nozadze[1,2]
[1]Institute of Chemical Biology, Ilia State University, Tbilisi, Republic of Georgia
[2]I. Beritashvili Center of Experimental Biomedicine, Tbilisi, Republic of Georgia

RESEARCH SUMMARY

Epilepsy is a heterogeneous group of disorders. It is the most common neurological disorder after stroke. Despite achieved progress in the treatment of epilepsy, about one third of the patients with epilepsy are resistant to existing pharmacotherapies. The current treatment of epilepsy focuses exclusively on preventing or suppressing seizures, which are the symptoms of the underlying disease.

The most important challenge is to prevent epileptogenesis, the process by which brain becomes epileptic. Drugs that prevent the development of epilepsy are not yet available.

The central nervous system is characterized with relatively high concentrations of myo-inositol (MI) as well as the means to synthesize it. MI serves not only as a precursor molecule for inositol lipid synthesis, but also as a physiologically important osmolyte. Alterations in MI deposition may playrole in a number ofneuropathological conditions, either as a physiologically important osmolyte or as a precursor molecule for phosphoinositide synthesis.

Several lines of evidence indicate the involvement of changes in inositols and other osmolytes in epilepsy and the effects of inositols in the regulation of induced seizures. We have revealed that the water extract of the medicinal plant *Aquilegia vulgaris* (a plant widely used in Oriental folk medicine as antiepileptic and soporific treatments) contains compounds altering binding of ligands to the benzodiazepine andgamma-aminobuturic acid (GABA) binding sites of the GABA-A receptors. We have identified two such compounds of this extract: (1) (MI) and (2) sleep-inducing lipid oleamide. Further we have shown that MI and scyllo-inositol (SCI) pretreatment significantly decreases the severity of seizures induced either by pentylentetrazolium or kainic acid (KA). As these effects were achieved by

This chapter was previously published in *Inositol: Synthesis, Functions and Clinical Implications*, edited by Henrique Rocha and Marina Cardoso, New York: Nova Publishers Inc., 2013.

physiological concentrations of the administered inositols, we have hypothesized that MI and SCI could represent endogenous anti-convulsants.

Our original data showed that MI also could interfere with the process of epileptogenesis. We have first induced status epilepticus by KA and then applied MI daily treatment. We have shown that a 28-day MI treatmentsignificantly attenuates biochemical changes associated with the process of epileptogenesis in the hippocampus, and restores the amount of some drastically reduced proteins to the normal level. To these proteins belongs the GLUR1 subunit of glutamate receptors, calcium/calmodulin dependent protein kinase II (CaMKII), a $\gamma 2$ subunit of GABA-A receptors. The obtained results indicate that MI treatment could at least modify the epileptogenesis process induced bybrain insult. Further studies of MI and SCI action could lead to more successful translational research and development of inositols as future anti-epileptic compounds.

In: Biochemistry Research Summaries. Volume 1
Editors: James T. Walker and Catherine M. Jones
ISBN: 978-1-53610-895-8

Chapter 103

THE CONTRIBUTION OF INOSITOL TRISPHOSPHATE SIGNALING TO CARDIAC PATHOLOGIES[#]

Elizabeth A. Woodcock
Molecular Cardiology Laboratory, Baker IDI Heart and Diabetes Institute, Melbourne, Victoria, Australia

RESEARCH SUMMARY

Inositol(1,4,5) trisphosphate is generated at the cell surface following stimulation and via activation of its receptors on intracellular stores, is the prime regulator of cellular Ca^{2+} responses. However, Ca^{2+} regulation in the heart depends most importantly on Ca^{2+}-induced Ca^{2+} release controlled by ryanodine receptors on the sarcoplamic reticulum, raising questions about the contribution of IP_3 and its receptors (IP_3-R). Furthermore, cardiomyocytes show only limited activity in generating IP_3 and the level of expression of IP_3-R is low. However, both IP_3 generation and IP_3-R expression are heightened in many cardiac pathologies, including hypertrophy, heart failure and in atrial fibrillation. IP_3 has been suggested to contribute to disease progression by promoting arrhythmia, hypertrophy and chamber dilatation. Experimental studies have argued both for and against this proposal. Many factors are altered in diseased myocardium and the challenge is to identify those that critically influence outcomes. In the current review, the evidence that IP_3 and IP_3-R contribute to heart disease is critically re-evaluated.

[#] This chapter was previously published in *Inositol: Synthesis, Functions and Clinical Implications*, edited by Henrique Rocha and Marina Cardoso, New York: Nova Publishers Inc., 2013.

In: Biochemistry Research Summaries. Volume 1
Editors: James T. Walker and Catherine M. Jones

ISBN: 978-1-53610-895-8

Chapter 104

THE ROLE OF INOSITOL 1,4,5-TRIPHOSPHATE RECEPTOR IN HEPATORENAL SYNDROME[#]

Dong-Lei Wang ***and Pei Liu***†

Department of Infectious Diseases, The First Affiliated Hospital,
China Medical University, Shenyang City, Liaoning Province, P. R. China

RESEARCH SUMMARY

Many hormone and neurotransmitter receptors utilize the phosphate-dylinositol signaling cascade. Phosphatidylinositol 4,5-bisphosphate is converted into two putative second messengers, inositol 1,4,5-trisphosphate (IP_3) and diacylglycerol. IP_3 regulates various kinds of cellular responses. It induces the release of Ca^{2+} from the intracellular stores by binding to the IP_3 receptor (IP_3R) on the endoplasmic reticulum. Many cell types, often nonexcitable, including glomerular mesangial cells (GMC) and renal vascular smooth muscle cells (VSMC), depend on this pathway to couple external signals to intracellular Ca^{2+} release. IP_3Rs not only triggers the release of calcium from intracellular stores but also opens plasma membrane calcium channels. Hepatorenal syndrome (HRS) is a common complication of advanced cirrhosis, characterized by renal failure and major disturbances in circulatory function. Renal failure is caused by intense vasoconstriction of the renal circulation.

HRS develops in approximately 55% of all patients with FHF. In order to delineate the role of inositol 1,4,5-triphosphate receptor in HRS, we investigated renal expression of type 1 inositol 1,4,5-trisphosphate receptors (IP_3R1) in mice with FHF. FHF model was induced by intravenous injection of lipopolysaccharide (LPS) in D-galactosamine (GalN)-sensitized mice via the caudal vein. GalN/LPS treatment led to a marked renal dysfunction and hyperkalaemia, of which kidney perfusion pressure was also significantly enhanced. However, it did not see any significant change in renal histopathology of rats in FHF. Simultaneously, the expression of protein and mRNA levels of renal IP_3R1 increased

---

[#] This chapter was previously published in *Inositol: Synthesis, Functions and Clinical Implications*, edited by Henrique Rocha and Marina Cardoso, New York: Nova Publishers Inc., 2013.

* Dong-Lei Wang. E-mail: wdl19851130@163.com.
† Pei Liu. E-mail: syliupei2013@163.com.cn.

markedly, serum TNF-α and endothelial-1(ET-1) levels reached a maximum level. We employed an inhibitor of IP_3Rs (2-APB) in this renal failure model, and found that reduction in the glomerular filtration rate (GFR) was reversed and hyperkalaemia recovered after administration of 2-APB; whereas, 2-APB did not affect the liver injury and the expression of IP_3R1. In contrast, intravenous injection of the TNF-α monoclonal antibody into the caudal vein significantly decreased the mortality rate of GalN/LPS-treated rats from 70% to zero, and significantly improved the compromised GFR. In TNF-α monoclonal antibody treated rats, serum TNF-α, ET-1 levels and expression of IP_3R1 were decreased. We had demonstrated in a recent study that TNF-α could enhance IP_3R1 expression at mRNA and protein levels in human mesangial cells (HMCs), mediated through TNFR1/PC-PLC/PKCα-dependent and TNFR2-dependent IP_3R1 pathways. Isolated perfusion kidney experiments demonstrated that in normal rat kidney preperfused with TNF-α, stimulation with ET-1 could raise the perfusion pressure evidently and overexpress IP_3R1 protein and mRNA levels. Similar results were seen in GMCs and VSMCs incubated by TNF-α, the calcium release markedly increased after treated with ET-1. The findings in the present study demonstrated that TNF-α-dependent IP_3R1 up-regulation might be a specific target in future studies of renal dysfunction during the end-stage of liver disease.

In: Biochemistry Research Summaries. Volume 1
Editors: James T. Walker and Catherine M. Jones

ISBN: 978-1-53610-895-8

Chapter 105

PATHOPHYSIOLOGIC SIGNIFICANCE OF INOSITOL PHOSPHATES (IPS) IN ALZHEIMER'S DISEASE#

Krishnaswamy Kannan[1], Rakhee Agarwal[2] and Nawab Ali[3,*]
[1]Department of Internal Medicine, Division of Rheumatology,
[2]Department of Pharmacology and Toxicology,
University of Arkansas for Medical Sciences, Little Rock, AR
[3]Department of Applied Science, University of Arkansas at Little Rock,
Little Rock, AR, US

RESEARCH SUMMARY

Alzheimer's disease (AD) is the leading cause of adult onset dementia and the incidence is expected to rise in US, Japan and other developed countries. AD is characterized by the accumulation of amyloid–beta peptides forming plaques. Though the neuropathology of AD was discovered over a century ago, the molecular aspects of the disease development are unclear. It is speculated that amyloid-beta peptides target cholinergic neurons leading to neural degeneration. Defective calcium signaling in neurons has been shown to play a critical role in the pathogenesis of AD. There is now increased interest and research on inositol phosphates (IPs) and calcium signaling since IPs are known to regulate Ca^{2+} homeostasis in endoplasmic reticulum, cell signaling, and cell death mechanisms. Thus, dysregulation of the IP mediated calcium homeostasis and associated cell physiology seems to play a pathological role in the development of many chronic illnesses particularly neurodegenerative diseases. In addition to IPs, aluminum toxicity also has been shown to contribute significantly to the perturbance of calcium metabolism, mitochondrial dysfunction, and inflammasome activation. There appears to be a convergence of dysregulated calcium metabolism, aluminum toxicity, and inflammasome activation leading to beta amyloid deposits and associated

This chapter was previously published in *Inositol: Synthesis, Functions and Clinical Implications*, edited by Henrique Rocha and Marina Cardoso, New York: Nova Publishers Inc., 2013.

* Corresponding author: Nawab Ali, PhD, Associate Professor, Department of Applied Science, University of Arkansas at Little Rock, 2801 S University Avenue, Little Rock, Arkansas 72204. Phone: 501-569 8003, Fax: 501-569 8020, E-mail: nali@ualr.edu.

pathology in Alzheimer's. Similar to Alzheimer's, IPs also have been shown to influence the pathogenesis of other neurological diseases including bipolar disorder. At the cellular level, there might be common shared pathways and underlying mechanisms related to IP. In this review, the pathophysiologic significance of IPs in the development of AD and bipolar disorders is discussed.

In: Biochemistry Research Summaries. Volume 1
Editors: James T. Walker and Catherine M. Jones
ISBN: 978-1-53610-895-8

Chapter 106

DESIGN AND SYNTHESIS OF BIOTINYLATED INOSITOL PHOSPHATES: APPLICATION TO THE INOSITOL PHOSPHATE-PROTEIN BINDING ANALYSIS[#]

Kensaku Anraku[1], Mikako Fujita[2], Takashi Morii[3], Yasuo Mori[4], Yoshinari Okamoto[5] and Masami Otsuka[5]
[1]Institute of Health Sciences, Kumamoto Health Science University, Izumi-machi, Kita-ku, Kumamoto, Japan
[2]Research Institute for Drug Discovery, School of Pharmacy, Kumamoto University, Oe-honmachi, Chuo-ku, Kumamoto, Japan
[3]Institute of Advanced Energy, Kyoto University, Uji, Kyoto, Japan
[4]Laboratory of Molecular Biology, Department of Synthetic Chemistry and Biological Chemistry, Graduate School of Engineering, Kyoto University, Kyoto, Japan
[5]Department of Bioorganic Medicinal Chemistry, Faculty of Life Sciences, Kumamoto University, Oe-honmachi, Chuo-ku, Kumamoto, Japan

RESEARCH SUMMARY

Inositol phosphates play important roles as second messengers in intracellular signal transduction. In order to study the relative affinity and specificity in binding inositol phosphates and diverse inositol phosphate-binding proteins by Surface Plasmon Resonance (SPR) analysis and pull-down analysis, biotinylated inositol phosphates, to be immobilized on avidin-based sensor chip or resin, were designed. The synthesis of the biotinylated inositol phosphates was accomplished by assembling the inositol phosphate building block, synthesized starting with optically resolved *myo*-inositol derivatives, and the biotin building block through a phosphate linkage. The inositol phosphate moiety of each biotinylated inositol phosphate showed specific binding to Pleckstrin Homology (PH) domain of PLCδ

[#] This chapter was previously published in *Inositol: Synthesis, Functions and Clinical Implications*, edited by Henrique Rocha and Marina Cardoso, New York: Nova Publishers Inc., 2013.

(phospholipase Cδ), PH domain of Grp1 (general receptor of phosphoinositides 1), and precursor of Gag protein ($Pr55^{Gag}$) of human immunodeficiency virus type 1 (HIV-1), as revealed by SPR analysis and pull-down analysis.

In: Biochemistry Research Summaries. Volume 1
Editors: James T. Walker and Catherine M. Jones
ISBN: 978-1-53610-895-8

Chapter 107

HEALTH BENEFITS OF INOSITOL AND INOSITOL HEXAPHOSPHATE[#]

Tzi Bun Ng and Charlene Cheuk Wing Ng*
School of Biomedical Sciences, Faculty of Medicine,
The Chinese University of Hong Kong, New Territories, Hong Kong, China

RESEARCH SUMMARY

Inositol hexaphosphate (InsP6, IP6, phytic acid) is found in abundance in beans and cereals. InsP6 occurs at much lower levels in most mammalian cells, in which they regulate cellular functions. Altered inositol production has been reported from patients with diabetes mellitus, chronic renal failure, galactosemia, and multiple sclerosis. IP6 has drawn much attention due to its activity in cancer prevention and control of experimental tumor growth, progression, and metastasis. InsP6 increases differentiation of malignant cells which may then revert to the normal phenotype. IP6 treatment of all the cell lines tested so far demonstrates that it is cytostatic and not cytotoxic. InsP6 inhibits growth and induces terminal differentiation of HT-29 human colon cancer cells. Studies of the expression of tumor suppressor gene demonstrate up-regulation of wild type p53 and down-regulation of the mutant form. Its efficacy in treating central nervous system disorders including Alzheimer's disease, depression, panic disorder, and obsessive-compulsive disorder has been reported. Its usefulness as an analgesic, in pediatric respiratory depression syndrome, to mitigate adverse effects associated with lithium treatment, and in preventing neural tube defects in embryonic mice has been documented. IP(6) exhibits immunoenhancing and hypocholesterolemic activities, minimizes pathological calcification and kidney stone formation, and lowers pathological platelet activity. However the risk of inducing uterine contractions restricts its usefulness in pregnancy.

[#] This chapter was previously published in *Inositol: Synthesis, Functions and Clinical Implications*, edited by Henrique Rocha and Marina Cardoso, New York: Nova Publishers Inc., 2013.
* Email: b021770@mailserv.cuhk.edu.hk, fax: 852 26035123, phone: 852 39436872.

In: Biochemistry Research Summaries. Volume 1
Editors: James T. Walker and Catherine M. Jones
ISBN: 978-1-53610-895-8

Chapter 108

CELLULAR FUNCTIONS OF INOSITOL[#]

Tzi Bun Ng and Charlene Cheuk Wing Ng*
School of Biomedical Sciences, Faculty of Medicine,
The Chinese University of Hong Kong, New Territories, Hong Kong,China

RESEARCH SUMMARY

Calcium-mobilizing receptors stimulate the hydrolysis of phosphatidylinositol 4,5-bisphosphate to 1,2-diacylglycerol and inositol 1,4,5-trisphosphate(Ins 1,4,5P3). Ins 1,4,5P3 released to the cytosol serves as a second messenger to release calcium from the endoplasmic reticulum. Growth factors such as platelet-derived growth factor, bombesin and vasopressin in fibroblasts or antigen in lymphocytes use the aforementioned transduction mechanism to produce intracellular mitogenic signals. The signaling mechanism via histamine H1-receptors is also used to mediate endothelial cell functions. Insulin regulates the phosphatidylinositol(PI) 3-kinase involving a physical association between the insulin receptor and the PI 3-kinase and tyrosine phosphorylation. Inositol is a nutrient required for keratinocyte proliferation. The Ins1,4,5P3/Ca2+ signal pathway functions to regulate such diverse processes as egg maturation and fertilization, growth, secretion, metabolism, neural activity, and excitation-contraction coupling in skeletal muscle. Myo-inositol serves as a clinically relevant osmolyte in the CNS, and its hexakisphosphate and pyrophosphorylated derivatives may play roles in such diverse cellular functions as DNA repair, nuclear RNA export and synaptic membrane trafficking. Glycerophosphoinositols induce cell proliferation in thyroid cells, modulate actin cytoskeleton organization in fibroblasts, and reduce the invasive potential of tumour cell lines. Glycerophosphoinositols promote cytokine-dependent chemotaxis in T-lymphocytes induced by SDF-1alpha-receptor activation.

[#] This chapter was previously published in *Inositol: Synthesis, Functions and Clinical Implications*, edited by Henrique Rocha and Marina Cardoso, New York: Nova Publishers Inc., 2013.
[*] Email:bo21770@mailserv.cuhk.edu.hk, fax;852 26035123, phone:852 39436872.

In: Biochemistry Research Summaries. Volume 1
Editors: James T. Walker and Catherine M. Jones

ISBN: 978-1-53610-895-8

Chapter 109

METABOLIC ASPECTS OF MACRONUTRIENTS[#]

Mostafa I. Waly, PhD
Department of Food Science and Nutrition, CAMS, Sultan Qaboos University, Muscat, Oman

RESEARCH SUMMARY

This book is designed and formatted for an undergraduate student level and fits with the curriculum of biochemistry courses in all medical universities throughout the world. This book represents an educational material which will provide students with a simple understanding of basic concepts of the macronutrient metabolism, biochemical events of macronutrients inside human cells, and cellular regulation of different metabolic pathways. In brief, this textbook provides insight into the nutritional biochemistry of macronutrients (carbohydrates, fat and protein) and their metabolic fate. The book covers basic concepts of digestion and absorption of three essential nutrients and their utilization by human cells in anabolic and catabolic pathways. The book is designed to link basic biochemistry to metabolism with a specific reference to the enzymatic, hormonal regulation and integrated metabolic pathways.

Special attention was given to the central role of glucose in controlling energy production and its relevance to enzyme activation and inhibition. The major pathways that regulate carbohydrates, fat and protein metabolites in different organs are discussed in correlation to the metabolic fate of each nutrient and the cellular responses to different hormones. The book provides a thorough understanding and lays the foundation for the metabolic basis of macronutrients and the biochemistry underlying the diseases pathogenesis. Finally the main theme of this book is to address the macronutrient metabolism in terms of anabolic pathways, catabolic pathways and amphibolic pathways. Anabolic pathways require energy which is provided by breaking down adenosine triphosphate (ATP) for the biosynthesis of protein, glycogen and triglycerides. Catabolic pathways involve oxidative processes which release free energy and the macronutrients are broken down in to their simple building blocks, e.g.,

---

[#] This chapter was previously published as a book: *Metabolic Aspects of Macronutrients*, by Mostafa I. Waly, New York: Nova Publishers Inc., 2013.

proteins are degraded into amino acids, polysaccharides are degraded into monosaccharides and triglycerides into to free fatty acids and glycerol.

In: Biochemistry Research Summaries. Volume 1
Editors: James T. Walker and Catherine M. Jones
ISBN: 978-1-53610-895-8

Chapter 110

β-GLUCANS AS NATURAL BIOLOGICAL RESPONSE MODIFIERS[#]

Vaclav Vetvicka
Department of Pathology, University of Louisville, Louisville, KY, US

RESEARCH SUMMARY

Natural products that can be used in prevention or treating various diseases have been sought throughout the history. One of the most promising, and at the same time most studied molecule, is β-glucan. The natural polysaccharides, usually termed "glucans," and referred to throughout this book, are non-cellulosic polymers of β-d-glucose, with glycosidic bonds in position β-(1-3) and usually with a certain, often high, portion of β-(1-6) bound glucose molecules. These molecules are isolated from numerous sources such as fungi, seaweed, bacteria, or grain. Due to the natural character, the continuing main problem in characterizing natural products also occurred here - natural substances usually represent a complex mixture of ingredients, each of which might significantly contribute to biological activity. The considerable heterogeneity of all natural glucans, regardless the source or methods of isolation, obviously was and continues to be the cause of a series of mutually contradicting conclusions, significantly diminishing the interest of pharmaceutical companies. As a result of this problem, the only meaningful data has come from experiments based on sufficiently purified and characterized glucan. Despite decades of intensive studies, our knowledge of glucans, their structural characteristics, biological activities, and their relationships, is still far from complete.

Most of the original studies of glucan effects focused on mice. Subsequently, numerous research showed that glucan has a similar, if not identical, immunomodulating and/or immunostimulating activity in a wide variety of other species, including shrimp, fish, rats, rabbits, dogs, chicken, pigs and horses. It is not surprising, therefore, that subsequently attention was also focused on humans. Based on all available data, it can been concluded that

---

[#] This chapter was previously published as a book: *β-Glucans as Natural Biological Response Modifiers*, by Vaclav Vetvicka, New York: Nova Publishers Inc., 2013.

glucan represents a natural immunomodulator that is active over the broadest spectrum of biological species and that it is one of the few immunostimulating agents spanning the entire evolutionary spectrum. Glucan can be considered to be an evolutionary old stimulant of immune responses. In addition, glucan remains similarly active regardless methods of administration, including intraperitoneal, intramuscular, subcutaneous, and intravenous applications. In the last decade, an interest in human applications has brought about studies of orally-administered glucan with results quite comparable to those achieved with injected glucan.

A chapter *Major Glucan Receptors* provides all the details regarding two major groups of glucan receptors—CR3 and Dectin-1. In addition, less studied receptors, such as langerin or lacrosylceramide, are described. The next chapter *Signaling Events upon Glucan-Cell Interaction* is focused on known mechanisms of glucan action, resulting from interaction with specific receptors and subsequent intracellular signaling. This information is followed by the chapter *Glucan and Bone Marrow* which details the extensive studies of how glucan mobilize bone marrow progenitors and stem cells into the peripheral organs resulting in using glucan as protection against negative side effects of irradiation or chemotherapy. These data are further explained in the chapter *Antitumor Activity of Glucan*, which is focused on current knowledge of mechanisms of glucan action in antitumor treatment.

The next part of the book is focused on the lesser known effects of glucan. First, a chapter *Glucan and Allergy* describes the possible effects in suppression of various allergic problems. Next, the chapter *Glucan in Aquaculture* deals with the role glucan plays in commercial farming of shrimp and fish. One of the newest effects of glucan is the suppression of various immunotoxicities, including mercury or perfluorinated compounds. These activities are given in the chapter *Glucan and Immunotoxicity*. Similarly, glucan has recently been found to help in the treatment of arthritis. Our current knowledge is summarized in the chapter *The Role of Glucan in Arthritis*. Greater attention has lately focused on the role glucan plays in modification of gastrointestinal problems, including colitis or inflammatory bowel disease. These effects are the subject of the chapter *Role of Glucan in Biology of Gastrointestinal Tract*. The role of glucan in suppression of various infections has been studied since the original findings of glucan activities, however, this subject has been abandoned lately. The chapter *Glucan and Infections* summarizes currently available information. The last two chapters, *Glucan: Other Biological Effects* and *Less Knows Effects of Glucan*, describe less known aspects of biological effects of glucan, from wound healing to lowering cholesterol levels or influencing the level of blood sugar.

In this book, we tried to offer information on both theoretical and practical aspects of glucan compiling both experimental and clinical studies that used both animal and human models. The book is for every reader seeking prompt and precise information on both history of glucan and on various biological and immunological effects including possible use in clinical practice. From the data summarized in this book it will be clear that more and more current data has brought additional insights into the multiple roles glucan plays in influencing various biological reactions. From the information available in the chapters mentioned above, a simple conclusion can be reached – glucan has significant potential in the activation of the relevant immune reactions in both animals and humans.

In: Biochemistry Research Summaries. Volume 1
Editors: James T. Walker and Catherine M. Jones

ISBN: 978-1-53610-895-8

Chapter 111

BIOLOGICAL ACTIVITY OF NATURAL PRODUCTS[#]

Jaroslava Švarc-Gajić
Department for Applied and Engineering Chemistry, Faculty of Technology,
University of Novi Sad, Novi Sad, Serbia

RESEARCH SUMMARY

Recognition that various naturally-occurring substances can affect different functions of man and other living organisms came long time ago. In order to exploit biological activity of both natural and synthetic compounds it is important to understand mechanisms underlying effects of different bioactive products.

The book deals with natural bioactive compounds of different origin, including plant, microbial, marine and animal chemicals. Material starts with explanation of bioactivity and basic toxicological principles that are necessary for understanding behavior of xenobiotics upon their entrance in human body. Since natural bioactive compounds were used for millennia by humans, some traditional and alternative medicinal approaches are explained in the book, including Ayurveda, Chinese medicine and homeopathy, as frequently used alternative medical practices today. Because natural sources of bioactive compounds are valuable industrial row materials, technological processes of isolation of several important compounds are elaborated step by step, explaining extraction, isomerization and purification stages at the industrial scale. As examples, technological production processes of ergot and opium alkaloids, as well as bioactive compounds from marine organisms are presented.

Phytochemicals are most readily available and assessable sources of natural bioactive compounds, so material covered by the book deals with some plant chemicals that may exhibit different biological effects, from beneficial to adverse. The use of plants and plant chemicals in traditional medicinal preparations, but also by a modern medicine, are explained.

Alkaloids fall into classes of natural most potent products. The use of ergot, purine, nux-vomica and other alkaloids by modern pharmacology is explained in the book, identifying natural sources and explaining mechanisms of biological effects.

[#] This chapter was previously published as a book: *Biological Activity of Natural Products*, by Jaroslava Švarc-Gajić, New York: Nova Publishers Inc., 2013.

Mostly beneficial health effects have been associated with natural phenols, however under certain circumstances these natural, plant-derived chemicals can exhibit adverse health effects in both humans and animals. After giving classification of natural phenolic compounds, book explains dual character of some natural phenols, focusing on gossypol, phlorizin, resveratrol, hydoxytyrosol, some phenolic acids and antraquinones. In addition, the use of some natural phenols in modern medical practice is explained.

Spices have long been used in different cultures as aphrodisiacs and remedies due to their potent bioactive chemicals. In some specific diets spices can represent an important source of bioactive chemicals and can provoke certain physiological effects. Some fruits, grains, legumes and vegetables also produce physiologically-active compounds and are more probable to provoke biological response due to greater quantities of these dietary products used in comparison to spices.

Today venoms are used as valuable medicinal compounds for both diagnostic and curative purposes. For diagnosis of blood chemistry deficiencies snake venoms are commonly used, whereas improved, non-addictive analgesics are based on cone snail venoms. The book stresses medicinal usability of venoms, produced by different organisms, including insects, marine species, snakes, lizards etc. In addition, biochemistry, production and use of antivenoms are explained.

Chemometry is multidisciplinary scientific discipline heavily used in metabolomics, analytical chemistry, biochemistry, medicine etc. This modern science may be very useful in predicting biological effects of natural products, as well as their derivatives. Since natural products are great inspiration to modern pharmacology, chemometric analysis is indispensable in studying natural bioactive compounds. The book explains basic chemometric principles and most frequently used chemometric models, including QSPR (Quantitative Structure Property Relationship), QSAR (Quantitative Structure Activity Relationship) and QSRR (Quantitative Structure Retention Relationship). Further, the material deals with *in silico* modeling that is used in modern science to define or predict molecular properties or to predict bioactivity, basic pharmacokinetic steps and interaction with receptors.

In: Biochemistry Research Summaries. Volume 1
Editors: James T. Walker and Catherine M. Jones
ISBN: 978-1-53610-895-8

Chapter 112

FATTY ACIDS: TYPES, ROLES AND HEALTH BENEFITS[#]

Narcisa Maria Bandarra*[*]*, Irineu Batista and Maria Leonor Nunes
INRB/IPIMAR
Av. Brasília Lisboa, Portugal

RESEARCH SUMMARY

Fatty acids (FA) are organic molecules constituted by an alquilic chain with a carboxylic group at extremity. These compounds are mainly found as part of the basic structure of triacylglycerols (TAG), phospholipids (PL) and cholesterol esters (CE) and as free fatty acids (FFA).FA classification is based according to the length and the degree of unsaturation of the carbon chain. Generally are constituted by an even number of carbon atoms and a chain length of 14-24 carbons. Animals FA usually present linear chains and a number of carbon atoms between 14 and 22 but on occasion can span the range from 2 to 36 or more. According to the chemical bonds of the fatty acid structure, they can be categorized as saturated fatty acids (SFA), without any double bonds in the structure, monounsaturated fatty acids (MUFA) with one double bond and polyunsaturated fatty acids (PUFA) with two or more double bonds in a methylene interrupted position.

[#] This chapter was previously published in *Fatty Acids: Types, Roles and Health Effects*, edited by Grace Chen, New York: Nova Publishers Inc., 2013.

[*] E-mail: narcisa@ipimar.pt

In: Biochemistry Research Summaries. Volume 1
Editors: James T. Walker and Catherine M. Jones

ISBN: 978-1-53610-895-8

Chapter 113

INVOLVEMENT OF N-3 POLYUNSATURATED FATTY ACIDS IN BRAIN RESISTANCE TO AGEING: EXPERIMENTAL EVIDENCE AND UNDERLYING MECHANISMS#

I. Denis*[*1], B. Potier[2], S. Vancassel[1], C. Heberden[1] and M. Lavialle[1]
[1]Unité de Nutrition et régulation lipidique des fonctions cérébrales (NuReLiCe), INRA. Jouy-en-Josas cedex, France
[2]INSERM U, Centre de psychiatrie et neurosciences, Paris, France

RESEARCH SUMMARY

The increasing life expectancy in the populations of rich countries raises the pressing question of how the elderly can maintain their cognitive function. Cognitive decline is characterized by the loss of short-term memory due to a progressive impairment of the underlying brain cell processes. Age-related brain damage has many causes, some of which may be influenced by diet. An optimal diet may therefore be a practical way of delaying the onset of age-related cognitive decline. Nutritional investigations indicate that the n-3 poyunsaturated fatty acid (PUFA) content of western diets is too low to provide the brain with an optimal supply of docosahexaenoic acid (DHA), the main n-3 PUFA in cell membranes. Insufficient brain DHA has been associated with memory impairment, emotional disturbances and altered brain processes in rodents. Human studies suggest that an adequate dietary intake of n-3 PUFA can slow the age-related cognitive decline and may also protect against the risk of senile dementia. However, despite the many studies in this domain, the beneficial impact of n-3 PUFA on brain function has only recently been linked to specific mechanisms.

This chapter was previously published in *Fatty Acids: Types, Roles and Health Effects*, edited by Grace Chen, New York: Nova Publishers Inc., 2013.

* Corresponding author: Isabelle Denis, Unité NuReLiCe – INRA – 78 352 Jouy-en-josas cedex 2 – France. isabelle.denis@jouy.inra.fr. tel: 33 (1) 34 65 23 13. fax : 33 (1) 34 65 23 11

This review examines the hypothesis that an optimal brain DHA status, conferred by an adequate n-3 PUFA intake, limits age-related brain damage by optimizing endogenous brain repair mechanisms. Our analysis of the abundant literature indicates that an adequate amount of DHA in the brain may limit the impact of stress, an important age-aggravating factor, and influences the neuronal and astroglial functions that govern and protect synaptic transmission. This transmission, particularly glutamatergic neurotransmission in the hippocampus, underlies memory formation. The brain DHA status also influences neurogenesis, nested in the hippocampus, which helps maintain cognitive function throughout life.

Although there are still gaps in our knowledge of the way n-3 PUFAs act, the mechanistic studies reviewed here indicate that n-3 PUFA may be a promising tool for preventing age-related brain deterioration.

In: Biochemistry Research Summaries. Volume 1
Editors: James T. Walker and Catherine M. Jones
ISBN: 978-1-53610-895-8

Chapter 114

ESSENTIAL FATTY ACIDS AND DISEASE STATES IN DOMESTIC ANIMALS[#]

Julia Schumann* and Herbert Fuhrmann
Institute of Physiological Chemistry, Faculty of Veterinary Medicine,
University of Leipzig, Leipzig, Germany

RESEARCH SUMMARY

The role of dietary fat in the etiology of chronic diseases, including cardiovascular disease, diabetes, obesity, cancer, arthritis as well as inflammatory disorders has been extensively studied. There is overwhelming evidence for the great importance of the type and the total amount of fat ingested on the course of numerous disorders. Remarkably, fat intake especially in combination with other dietary factors (e.g. antioxidants, micronutrients or method of food preparation) may either enhance or prevent disease processes.

Animal feeding studies have played a considerable role in establishing the dietary essentiality of linoleic acid (LA, C18:2n6) and alpha-linolenic acid (LNA, C18:3n3) and in elucidating the biochemical and physiological roles of arachidonic acid (AA, C20:4n6) and docosahexaenoic acid (DHA, C22:6n3). Of particular importance, animal studies offer the opportunity to strictly control the fatty acid intake as well as to explore any adverse effects of oils high in one fatty acid family and the potential for avoiding these effects by co-supplementation with fatty acids from another family. However, care must be taken not only to use the appropriate animal model but also to feed an adequate amount of fatty acids. Moreover, when exploring the gained information to humans one must consider likely species differences in growth, development, nutrient metabolism and requirements.

This article reviews studies in domestic animals making major contributions to the understanding of the role of dietary fatty acids for distinct developmental and disease states.

This chapter was previously published in *Fatty Acids: Types, Roles and Health Effects*, edited by Grace Chen, New York: Nova Publishers Inc., 2013.

* Correspondence: Dr. Julia Schumann, Phone: ++49-341-9738107, E-mail address: julia.schumann@vmf.uni-leipzig.de; Institute of Physiological Chemistry, Faculty of Veterinary Medicine, University of Leipzig, An den Tierkliniken 1, 04103 Leipzig, Germany

Particular attention is paid to lipid metabolic disorders, lactation as well as the function of the skin, the brain and the immune system. The review closes with suggestions concerning future research.

In: Biochemistry Research Summaries. Volume 1
Editors: James T. Walker and Catherine M. Jones
ISBN: 978-1-53610-895-8

Chapter 115

EFFECT OF FATTY ACID SYNTHASE ON FATTY ACIDS BIOGENESIS IN METABOLIC SYNDROME[#]

Yang Liu, Fang Zhang, Deliang Chen, Wei Zhang, Liang Zhang and Wenjun Ding*
College of Life Sciences,
Graduate University of the Chinese Academy of Sciences, Beijing, China

RESEARCH SUMMARY

Metabolic syndrome (MS) is characterized by a group of metabolic risk factors in one person. The metabolic risk factors include abdominal obesity, insulin resistance, atherogenic dyslipidemia, hyperlipidemia, hypertension, proinflammatory state and prothrombotic state. People with the metabolic syndrome are at increased risk of coronary heart disease, stroke, peripheral vascular disease and Type 2 diabetes mellitus. Accumulating evidence indicate that lipid metabolism is important to diabetes. Fatty acids have been shown to be involved in the development of insulin resistance associated with obesity. It has been demonstrated that fatty acid synthase (FAS, EC 2.3.1.85) is a multi-enzyme that plays a key role in fatty acid synthesis. Fatty acid synthase performs the anabolic conversion of dietary carbohydrate or protein to fatty acids. FAS catalyzes the synthesis of long chain fatty acids from acetyl-CoA and malonyl-CoA, such as palmitates. Another evidences show that increased circulating FAS levels is associated with metabolic disorders. Thus, FAS has been a potential marker for metabolic stress. Moreover, the effects of FAS on fatty acids biogenesis in metabolic syndrome are reviewed.

[#] This chapter was previously published in *Fatty Acids: Types, Roles and Health Effects*, edited by Grace Chen, New York: Nova Publishers Inc., 2013.

* Corresponding author: Tel: +8610-88256290; Fax: +8610-88256460. E-mail address: dingwj@gucas.ac.cn

In: Biochemistry Research Summaries. Volume 1
Editors: James T. Walker and Catherine M. Jones

ISBN: 978-1-53610-895-8

Chapter 116

DEVELOPMENTAL REGULATION OF HYDROXY FATTY ACID SYNTHESIS IN *LESQUERELLA FENDLERI*[#]

Grace Q. Chen[*]
Western Regional Research Center, Agricultural Research Service,
U.S. Department of Agriculture, Albany, CA, US

RESEARCH SUMMARY

A conventional source of hydroxyl fatty acid (HFA) is from castor oil which contains 90% ricinoleic acid (12-hydroxy-octadec-*cis*-9-enoic acid: 18:1OH). Ricinoleic acid and its derivatives are used as raw materials for numerous industrial products. The production of castor oil, however, is hampered by the presence of the toxin ricin and hyper-allergic 2S albumins in castor seed. *Lesquerella fendleri* (A. Gray) S. Wats. (Brassicaceae), being developed as a new industrial oilseed crop in the southwestern region of the United States, is valued for HFA. The majority of HFA in *L. fendleri* is lesquerolic acid (14-hydroxy-eicos-*cis*-11-enoic acid: 20:1OH) comprising 55-60% of total seed fatty acids. Lesquerolic acid is derived from a 2 carbon elongation of 18:1OH; lesquerella oil represents an alternative to castor oil as a safe source of HFA.

To advance our understanding of regulatory mechanisms underlying synthesis and accumulation of HFAs, we illustrated a wide range of morphological, physiological and biochemical changes that occur during seed development in *L. fendleri*. We also examined the expression patterns of two key genes involved in fatty acid synthesis, as well as the temporal changes of fatty acid composition during seed development. All of these measurements are in parallel with the same time course, it allowed us to investigate the relationships among HFA accumulation, gene expression and seed development.

[#] This chapter was previously published in *Fatty Acids: Types, Roles and Health Effects*, edited by Grace Chen, New York: Nova Publishers Inc., 2013.

[*] Corresponding author address: Grace Q. Chen, 800 Buchanan St., Albany, CA 94710, USA. Tel: 510-559-5627. Fax: 510-559-5768. Email, grace.chen@ars.usda.gov

In: Biochemistry Research Summaries. Volume 1
Editors: James T. Walker and Catherine M. Jones
ISBN: 978-1-53610-895-8

Chapter 117

BETA-CELL, LIPOTOXICITY AND OXIDATIVE STRESS[#]

Roberto Lupi*[*1] and Silvia Del Guerra[2]
[1]Department of Endocrinology and Kidney, Metabolic Unit "Renzo Navalesi", Azienda Ospedaliero-Universitaria Pisana, Pisa, Italy
[2]Department of Endocrinology and Metabolism, Metabolic Unit "Renzo Navalesi", University of Pisa, Pisa, Italy

RESEARCH SUMMARY

Under physiological conditions, free fatty acids (FFAs) sustain basal insulin secretion in the fasted state and potentiate hormone release in response to glucose. In contrast, prolonged exposure to increased FFA concentrations, as to occur in diabetes and obesity, has detrimental effects on pancreatic beta-cells, a phenomenon termed lipotoxicity. During few years, several studies have consistently reported that increased concentrations of FFA induce insulin secretion dysfunction and increased apoptosis in human beta-cells. These deleterious effects seem to be due more to saturated than unsaturated fatty acids, with these latter that may actually have a protective action.

It's well known that multiple mechanisms concur in generating the damage, including accumulation of malonyl CoA and long-chain fatty acyl CoA, increased fatty acid oxidation and esterification, accumulation of ceramide and fatty-acid induced apoptosis, but activation of oxidative stress may contribute to the final toxic effect. Hereby we review the available in-vitro data on lipotoxicity on human pancreatic beta-cells, and the evidence that reducing oxidative stress in islet cell exposed to FFA seems to be an effective method to improve beta-cell function and survival, suggesting that the antioxidant treatment may protect human islet beta-cell against injuries of lipotoxicity.

This chapter was previously published in *Fatty Acids: Types, Roles and Health Effects*, edited by Grace Chen, New York: Nova Publishers Inc., 2013.

* Correspondence to: Dr. Roberto Lupi, BHS. Departement of Endocrinology and Kidney, Metabolic Unit "Renzo Navalesi", Azienda Ospedaliero-Universitaria Pisana, 56124 Pisa – Italy. e-mail: lupir@immr.med.unipi.it. tel: +39050995134. fax: +39050541521

In: Biochemistry Research Summaries. Volume 1
Editors: James T. Walker and Catherine M. Jones
ISBN: 978-1-53610-895-8

Chapter 118

MICROBIAL PRODUCT MEDIATED IN SITU TRANSFORMATION OF VEGETABLE OILS TO FATTY ACID VESICLES[#]

Arumugam Gnanamani*[*]*, Varadharajan Kavitha and Ganesan Sekaran
Microbiology Division, Central Leather Research Institute, (CSIR, New Delhi),
Adyar, Tamil Nadu, India

RESEARCH SUMMARY

A question on 'how the life is originated?' was approached by number of researchers from time immemorial and the answers received from them were neither acceptable nor denied. However, a semisynthetic approach by a group of synthetic biologists found suitable to the existing living systems. The eminent synthetic biologists, Walde, Luisi, Szostak, Deamer, Dworkin reported 'Vesicles' are the key structures for origin of life based on the information provided by Morowitz, where, the author revealed lipid vesicle play a key role in the emergence of first living cells. According to synthetic biologists, vesicles are nothing but supramolecular assembly of bilayer membrane of macromolecules. Followed by them, number of researchers currently working in the field of vesicles confirmed and agreed. Vesicles formed naturally inside the cell are permeable and trap or transport substances of interest. This property was claimed for the vesicle as the source material for origin of life. Further, intensive research on vesicles by Szostak, Deamer Luisi and co-workers provided a plausible model based on the semi synthetic approach for the development of live cells from non-living components. In brief, their approach emphasizes the important key function of vesicles, i.e., trapping and transportation, may be used for the construction of artificial cells

[#] This chapter was previously published in *Fatty Acids: Types, Roles and Health Effects*, edited by Grace Chen, New York: Nova Publishers Inc., 2013.

[*] Author for Correspondence: Dr. A. Gnanamani, Scientist and Head, Microbiology Division, Central Leather Research Institute (CLRI), Adyar, Chennai 20, Tamil Nadu, India, Email: agmani_2000@yahoo.com, gnanamani3@gmail.com

by entangling the necessary components of a life, viz., DNA, enzymes, ribosomes, RNA, amino acids, NTPS.

In: Biochemistry Research Summaries. Volume 1
Editors: James T. Walker and Catherine M. Jones
ISBN: 978-1-53610-895-8

Chapter 119

PLANT-DERIVED BIOFUNGICIDES: OVERVIEW AND PERSPECTIVES[#]

Eva Švecová and Paola Crinò*
ENEA C.R. Casaccia, Unit of Sustainable Development
and Innovation of the Agro-Industrial System, Roma, Italy

RESEARCH SUMMARY

Natural products such as the plant-derived ones, e.g. plant extracts and essential oils, as tools of biological control have recently acquired a great scientific and commercial interest. This is due to environmental toxicity problems such as water contamination, persistence of residues in the plant, and to resistance of fungi to chemical fungicides. Many of these plant-derived products are used in the traditional medicine and often they are supposed to be efficient also in plant disease control. In fact, historically, plant products have a determinant role in agriculture and food preservation. The sources of these biofungicides are often plant secondary metabolites, which evolved in plants to confer a protection against microbial attack. Plant-derived biofungicide formulations may be represented by a single active compound, simple mixtures of natural ingredients or complex mixtures with multiple effects on the target pathogens. Such complex mixtures might act also as plant biostimulants. Due to the rich chemical composition of natural substances, it is difficult to identify exactly which component has the antifungal effect. In addition, many compounds act in synergy against different fungal species. The novel active compounds are continuously being discovered, thus responding to the consumer demand on the significant reduction of chemical usage in agriculture. Plant-derived biofungicides are essential to develop new biological control strategies required for the growing organic market. A case study on antifungal activity of *Vitex agnus-castus* extract against *Pythium ultimum* in tomato under *in vitro* and *in vivo*

[#] This chapter was previously published in *Fungicides: Classification, Role in Disease Management and Toxicity Effects*, edited by Miguel N. Wheeler and Bradley R. Johnston, New York: Nova Publishers Inc., 2013.

* Corresponding author: paola.crino@enea.it.

conditions is presented. In addition, the ability of *V. agnus-castus* extract to enhance plant defence responses connected with *PR* gene induction upon pathogen inoculation is discussed.

In: Biochemistry Research Summaries. Volume 1
Editors: James T. Walker and Catherine M. Jones

ISBN: 978-1-53610-895-8

Chapter 120

REVIEW OF THEIR ACTIVITY AND PATHOGEN RESPONSES[#]

Andrea Kunova and Paolo Cortesi*
Università degli Studi di Milano, Department of Food,
Environmental and Nutritional Sciences, Milano, Italy

RESEARCH SUMMARY

Magnaporthe oryzae, the causal agent of the rice blast disease, is one of the most important pathogens of cultivated rice worldwide. It can cause serious epidemics resulting in substantial yield and economic losses. Rice blast management relies on the use of fungicides in many regions. At present, there are approximately thirty fungicides registered worldwide, most of which are being used in Asia. In Europe, azoxystrobin and tricyclazole were registered for application on rice at the beginning of 1990's and in Italy are being extensively used since then. In this chapter, the main chemical classes of fungicides used in rice blast management will be characterized, with emphasis on azoxystrobin and tricyclazole, which will be described in more detail.

The molecular basis of their modes of action and their activity on various stages of pathogen life cycle will be defined. Despite having an important role in rice disease management, only scarce information about responses of the *M. oryzae* isolated from rice to azoxystrobin and tricyclazole are available, therefore the baseline response of the pathogen, risks and possible emergence of resistance to both fungicides will be discussed.

[#] This chapter was previously published *in Fungicides: Classification, Role in Disease Management and Toxicity Effects*, edited by Miguel N. Wheeler and Bradley R. Johnston, New York: Nova Publishers Inc., 2013.

* Corresponding author: Paolo Cortesi, Università degli Studi di Milano, Department of Food, Environmental and Nutritional Sciences, Via Celoria 2, 20133 Milano, Italy. E-mail: paolo. cortesi@unimi.it.

In: Biochemistry Research Summaries. Volume 1
Editors: James T. Walker and Catherine M. Jones
ISBN: 978-1-53610-895-8

Chapter 121

FUNGICIDES AND THEIR ROLE IN DISEASE MANAGEMENT[#]

Agostino Santomauro*[*] *and Franco Faretra
Department of Plant, Soil, and Food Sciences,
University of Bari "Aldo Moro", Bari, Italy

RESEARCH SUMMARY

Fungal diseases are known to affect produce quality on most crops and to cause severe economic losses worldwide. The use of fungicides has been for a long time the main tool to control many economically important fungal pathogens. In the last decades, increased public concern on food safety and environmental impact of farming activities has led regulatory Authorities to adopt very stringent rules on the evaluation and authorization of plant protection products (PPPs), especially for toxicological and environmental negative side effects related to their use. As a result, the development of new PPPs has been addressed at reducing their toxicity towards humans, mammalians and non-target species. In this framework, agrochemical firms have made available several new fungicides with novel specific (single-site) modes of action. Many of them have penetrant properties, are very selective, showing specific activity against target pathogens, as well as enhanced efficacy as compared to traditional multisite fungicides (i.e., carbamates, copper compounds, dithiocarbamates, quinones, sulphur, thiophtalimides, etc.). On the other hand, their single-site modes of action make them at moderate to high risk of inducing fungicide resistance in target pathogens and, hence, anti-resistance strategies must be carefully implemented in their use. Nowadays, about 200 active substances are available, with almost 40 different known specific target sites, plus a number of chemicals having multi-site or unknown mode of action. The increased availability of fungicides allows a more flexible and effective planning of crop protection strategies well fitting to Integrated Pest Management (IPM) principles. Yet,

---

[#] This chapter was previously published *in Fungicides: Classification, Role in Disease Management and Toxicity Effects*, edited by Miguel N. Wheeler and Bradley R. Johnston, New York: Nova Publishers Inc., 2013.

[*] Corresponding Author: agostino.santomauro@uniba.it.

it requires a constant update of technical and scientific expertise by farmers and advisors. This review focuses on the main features of the most commonly used fungicides and their role in disease control in the overall approach to IPM aiming at improving cropping sustainability.

In: Biochemistry Research Summaries. Volume 1
Editors: James T. Walker and Catherine M. Jones
ISBN: 978-1-53610-895-8

Chapter 122

CLASSIFICATION OF FUNGICIDES#

Erlei Melo Reis[1] and Marcelo Aníbal Carmona[2]
[1]Faculty of Agronomy and Veterinary Medicine, University of Passo Fundo, Passo Fundo, Brazil
[2]Faculty of Agronomy, University of Buenos Aires, Buenos Aires, Argentina

RESEARCH SUMMARY

Fungicides are chemicals agents that inhibit or eliminate the mycelial growth or fungal spores. The chemical, physical, and biological characteristics of a fungicide determines its suitability for control of a determined disease. Fungicides can be classified in different ways as fungistactic, anti-sporulant, disinfectant, preventative, protectant, residual, contact, curative, eradicative, topic, deep penetrating, non-penetrating, mesostemic, and systemic, among others. As a result, the literature is confusing to clearly understand and to discuss topics on fungicides. This chapter provides information regarding fungicides classification covering mainly fungicides used in aerial organs of plants.

This chapter was previously published *in Fungicides: Classification, Role in Disease Management and Toxicity Effects*, edited by Miguel N. Wheeler and Bradley R. Johnston, New York: Nova Publishers Inc., 2013.

In: Biochemistry Research Summaries. Volume 1
Editors: James T. Walker and Catherine M. Jones
ISBN: 978-1-53610-895-8

Chapter 123

EFFECTIVE FUNGICIDES FOR CEREAL CROPS PROTECTION AGAINST TOXICOGENIC FUNGI CAUSING *FUSARIUM* HEAD BLIGHT[#]

L. D. Grishechkina and V. I. Dolzhenko
All Russia Plant Protection Research Institute, St.-Petersburg, Russia

RESEARCH SUMMARY

Analyses the long-term effectiveness studies of novel fungicides use for *Fusarium* head blight control on winter wheat in Krasnodar territory of Russia are presented. Effects of fungicides on the grain infestation by the pathogen including latent infection are described. The hazard of studied preparations to the grain agrobiocenosis is shown.

[#] This chapter was previously published *in Fungicides: Classification, Role in Disease Management and Toxicity Effects*, edited by Miguel N. Wheeler and Bradley R. Johnston, New York: Nova Publishers Inc., 2013.

ng cellulose acetate in
arity in the agricultural
euticals such as aspirin
he chemical properties,

al use of acetic acid or
st two decades, as well

ISBN: 978-1-53610-895-8

JSTRIAL USE
CID[#]

a M. Mahungu[2,†]
Iuman Nutrition and Dietetics,

of Dairy and Food
erton, Kenya
ns, Nairobi, Kenya

IARY

hanoic acid is a carboxylic compound
' vinegar, which means "sour wine" in
e action of aerobic bacteria on dilute
his chemical change is the reaction of
negar is about 4-18 % acetic acid. The
Pure acetic acid is classified as a weak
or, with a freezing point below 16 °C

iethyl ether due to hydrogen bonding
red a variety of uses industrially and
eral deposits and organic compounds,
eservative, and disinfectant in cleaning

Properties, Production and Applications, edited

man Nutrition and Dietetics, P.O Box 45240-
P.O Box 36374, 00200 Nairobi, Kenya. E-mail:

d Food Science and Technology, P.O Box 536,

It is also used in industrial manufacture of acetate esters incluc textile industry and vinyl acetate. Use of acetic acid is gaining popu sector as a fungicide and herbicide and in the production of pharma using acetic anhydride. The aim of this chapter is to give insights into production and applications of acetic acid.

We will focus on the developments in the chemistry and industr vinegar (acetic acid and vinegar will be used interchangeably) in the as draw attention to new emerging areas in acetic acid research.

In: Biochemistry Research Summaries. Volume 1
Editors: James T. Walker and Catherine M. Jones
ISBN: 978-1-53610-895-8

Chapter 125

ACETIC ACIDS: INDOLE-3-ACETIC ACID (IAA), INDOLE-3-BUTYRIC ACID (IBA), A-NAPHTHALENEACETIC ACID (A-NAA) AND 2,4 DICHLOROPHENOXY ACETIC ACID (2,4-D): CHEMICAL PROPERTIES, PRODUCTION AND APPLICATIONS IN TISSUE CULTURE, ON ROOTING AND ON BIOCHEMICAL PARAMETERS[#]

Virginia Sarropoulou,[*] Kortessa Dimassi-Theriou and Ioannis Therios
Laboratory of Pomology, Department of Horticulture, School of Agriculture, Aristotle University of Thessaloniki, Thessaloniki, Greece

RESEARCH SUMMARY

In this chapter, the effects of the auxins IAA, IBA, a-NAA and 2,4-D from the category of the acetic acids in tissue culture, on rooting, on photosynthetic pigments (total leaf chlorophyll (a+b) and carotenoids content) as well as on total carbohydrate and proline concentration were analysed. Acetic acid is a weak organic compound with the chemical formula CH_3CO_2H (also written as CH_3COOH or $C_2H_4O_2$) and is the main component of vinegar. Acetic acid is mildly corrosive to metals including iron, magnesium, and zinc, forming hydrogen gas and salts called acetates. At physiological pHs, acetic acid is usually fully ionized to acetate. In biochemistry, acetate and acetic acid are equivalent. Acetic acid is produced industrially both synthetically and by bacterial fermentation. Glacial acetic acid is an excellent polar protic solvent, is often used as a solvent for reactions involving

---

[#] This chapter was previously published in *Acetic Acids: Chemical Properties, Production and Applications*, edited by Angelo Basile, New York: Nova Publishers Inc., 2013.

[*] Corresponding author: Virginia Sarropoulou. E-mail: vsarrop@gmail.com; vsarropo@agro.auth.gr.

carbocations, such as Friedel-Crafts alkylation and is used in analytical chemistry for the estimation of weakly alkaline substances such as organic amides. Auxins derive their name from the Greek word *auxein* - "to grow/increase".

The most important member of the auxin family is indole-3-acetic acid (IAA). Synthetic auxin analogs include 1-naphthaleneacetic acid, 2,4-dichlorophenoxyacetic acid (2,4-D), and many others. Auxins are also often used to promote initiation of adventitious roots, to promote uniform flowering and fruit set, and to prevent premature fruit drop. On the cellular level, auxin is essential for cell growth, affecting both cell division and cellular expansion. Auxin participates in phototropism, geotropism, hydrotropism and other developmental changes. Auxins promote root initiation, induce shoot apical dominance and delay fruit senescence. In low concentrations, auxins can inhibit ethylene formation and transport in plants; however, high concentrations can induce the synthesis of ethylene. IAA is chemically similar to the amino acid tryptophan which is generally accepted to be the molecule from which IAA is derived. 2,4-D is absorbed through the leaves and is translocated to the meristems of the plant. Auxins and their synthetic analogues are commonly used for rooting of cuttings, but their efficiency depends on experimental set-up and, even more importantly, on species or cultivar, and type of explants investigated. A favourable rooting effect of IBA may be due to its higher stability in comparison to IAA, its slow transport from the site of application at the base of cutting and its conversion to IAA in the cutting.

In: Biochemistry Research Summaries. Volume 1
Editors: James T. Walker and Catherine M. Jones
ISBN: 978-1-53610-895-8

Chapter 126

STRUCTURAL AND VIBRATIONAL PROPERTIES OF SOME QUINOLINE ACETIC ACID DERIVATIVES WITH POTENTIAL BIOLOGICAL ACTIVITIES[#]

E. Romano,[1] J. L. Pergomet,[2] J. Zinczuk[2] and S. A. Brandán[1,*]
[1]Cátedra de Química General. Instituto de Química Inorgánica.
Facultad de Bioquímica, Química y Farmacia. Universidad Nacional
de Tucumán. San Miguel de Tucumán. Tucumán, Argentina
[2]Instituto de Química Rosario (CONICET-UNR), Facultad de Ciencias
Bioquímicas y Farmacéuticas, Rosario, Santa Fé, Argentina

RESEARCH SUMMARY

Quinoline compounds are widely used in medicine due to their great variety of biological and biochemical properties, especially anti-malarial and anti-microbial activities. Recently, they have been studied for their modes of function in the inhibition of tyrosine kinases, proteasome, tubulin polymerization and DNA repair. Thus, certain quinoline derivatives also show effective anticancer activity. For these reasons, the quinoline derivatives have been the object of numerous studies from different points of view. In this chapter, a complete characterization of the synthesis, structural and vibrational properties of three quinoline acetic acid derivatives with potential biological activities are studied. The knowledge of the structural and vibrational properties is important for the design and synthesis of new and better quinolinic drugs.

In this chapter, the experimental and theoretical studies of the 2-(quinolin-8-yloxy), 2-(5-chloroquinolin-8-yloxy) and 2-(5,7-dichloroquinolin-8-yloxy) acetic acids are reported. The compounds were characterized by infrared and Raman spectroscopies in the solid phase and

---

[#] This chapter was previously published in *Acetic Acids: Chemical Properties, Production and Applications*, edited by Angelo Basile, New York: Nova Publishers Inc., 2013.

[*] Corresponding author: S. A. Brandán. Cátedra de Química General. Instituto de Química Inorgánica. Facultad de Bioquímica, Química y Farmacia. Universidad Nacional de Tucumán. Ayacucho 471. 4000. San Miguel de Tucumán. Tucumán. R. Argentina. Tel.: +54-381-4247752; Fax: +54-381-4248169; E- mail: sbrandan@fbqf.unt.edu.ar.

by nuclear magnetic resonance (^{1}HNMR and ^{13}CNMR) spectroscopy in a $CDCl_3$ solution. The density functional theory (DFT) together with the 6-31G* and 6-311++G** basis sets were used to study the structural and vibrational properties. The harmonic vibrational wavenumbers for the optimized geometries were calculated at the same theory levels. For a complete assignment of all the observed bands in the vibrational spectra the DFT calculations were combined with Pulay's scaled quantum mechanical force field (SQMFF) methodology in order to fit the theoretical wavenumbers values to the experimental ones. The force constants for the three compounds were calculated and compared among them. The calculated chemical shifts of the ^{1}HNMR and ^{13}CNMR for the compounds were obtained by GIAO Method using the B3LYP/6-311++G** method. Furthermore, the characteristics of the electronic delocalization, the topological properties of the electronic charge density together with the highest occupied molecular orbital (HOMO)-lowest unoccupied molecular orbital (LUMO) energy gaps for the three compounds were studied.

In: Biochemistry Research Summaries. Volume 1
Editors: James T. Walker and Catherine M. Jones
ISBN: 978-1-53610-895-8

Chapter 127

USE OF ACETIC ACID AS AN ACETYLATING AGENT OF ALCOHOLS AND PHENOLS CATALYZED BY RECYCLABLE SOLID ACIDS[#]

***Luis R. Pizzio,** Gustavo P. Romanelli*[†] *and Mirta N. Blanco*[‡]**
Centro de Investigación y Desarrollo en Ciencias Aplicadas "Dr. Jorge J. Ronco" (CINDECA), Departamento de Química, Facultad de Ciencias Exactas, Universidad Nacional de La Plata-CCT La Plata, CONICET, 47 Nº 257, 1900 La Plata Argentina

RESEARCH SUMMARY

Acetic acid is one of the simplest carboxylic acid, and an important chemical and industrial chemical for different uses. It is employed, for example, in the production of cellulose and, in the food, diluted acetic acid acts as an acidity and as a condiment. The global demand of acetic acid is very important, around 6.5 million tonnes per year.

The acetylation reaction of organic compounds is one of the most frequently used transformations in organic synthesis. Among the various protected groups used for the hydroxyl function, the acetyl group is the most convenient in view of its easy introduction, being stable to the acidic reaction condition and also easily removable by mild alkaline hydrolysis. The most common and inexpensive reagent used to perform the acetylation is the acetic anhydride however, from an environmental point of view, a much convenient method for performing the acetylation is the use of acetic acid as acetylating agent, because only water is produced as by product, and it is a relative cheaper route for industrial purposes when compared to anhydrides or acid chlorides.

[#] This chapter was previously published in *Acetic Acids: Chemical Properties, Production and Applications*, edited by Angelo Basile, New York: Nova Publishers Inc., 2013.
* Luis R. Pizzio: Centro de Investigación y Desarrollo en Ciencias Aplicadas "Dr. Jorge J. Ronco" (CINDECA), Departamento de Química, Facultad de Ciencias Exactas, Universidad Nacional de La Plata-CCT La Plata, CONICET, 47 Nº 257, 1900 La Plata, Argentina. E-mail: lrpizzio@quimica.unlp.edu.ar.
[†] Gustavo P. Romanelli. E-mail: gpr@quimica.unlp.edu.ar.
[‡] Mirta N. Blanco. E-mail: mnblanco@quimica.unlp.edu.ar.

In this chapter we reviewed the performance of different recyclable solid acids synthesized in our laboratory as catalysts for the use of acetic acid as acetylating agent: 1) alcohols and phenols acetylation using borated zirconia; 2) alcohols and phenols acetylation using borated zirconia modified with ammonium metatungstate; 3) isoamyl alcohol acetylation using tungstophosphoric acid supported on zirconia; and 4) isoamyl alcohol acetylation using partially substituted Cs and K salts of tungstophosphoric acid.

The alcohols and phenols acetylation with the borated zirconia-based catalysts gave very good yields to the desired products with very high selectivity, though the alcohol and phenol structure is a key in the yield to the acetylation product. The zirconia-supported tungstophosphoric acid prepared by the incipient wetness impregnation led to obtain isoamyl acetate with high activity and selectivity, and similar results were obtained using the Cs and K tungstophosphates, observing a correlation with the acidic properties of the catalysts.

In: Biochemistry Research Summaries. Volume 1
Editors: James T. Walker and Catherine M. Jones
ISBN: 978-1-53610-895-8

Chapter 128

HYDROGEN PRODUCTION VIA ACETIC ACID STEAM REFORMING REACTION USING MEMBRANE REACTORS[#]

A. Iulianelli, S. Liguori, F. Dalena, and A. Basile*
CNR-ITM, c/o University of Calabria, Rende (CS), Italy

RESEARCH SUMMARY

In recent years, hydrogen production from biomass transformation has been very attractive. Particularly interesting are methods in which the biomass is converted into intermediate liquid bio-fuels such as pyrolysis oil, acetic acid or ethanol. Among various renewable sources, acetic acid may be easily produced by fermentation of biomass. Furthermore, acetic acid, unlike methanol and ethanol, is not inflammable and, as a consequence, it represents a safe hydrogen carrier.

In particular, acetic acid may be easily transformed into hydrogen with high selectivity over effective catalysts such as Ni–Co and Ru. However, to the best of our knowledge a restricted literature is noticeable on hydrogen production from acetic acid and only a few manuscripts deal with conventional reactors usually operating at temperature higher than 500 °C. In these studies, acetic acid steam reforming reaction is carried out using different catalysts and a hydrogen rich-gas stream is produced, containing secondary products such as CO_2, CO, CH_4, etc. Therefore, with the purpose of producing CO_x-free hydrogen for PEM fuel cell applications, the hydrogen rich-gas stream needs to be purified. This chapter is aimed to study the utilization of inorganic membrane reactor technology for producing CO_x-free hydrogen in only one step from acetic acid steam reforming instead of the conventional reactors followed by further stages for hydrogen purification such as water gas shift reactors (high and low temperature), partial oxidation reactors and pressure swing adsorption devices.

[#] This chapter was previously published in *Acetic Acids: Chemical Properties, Production and Applications*, edited by Angelo Basile, New York: Nova Publishers Inc., 2013.

* Corresponding author: A. Basile. CNR-ITM, c/o University of Calabria, Via Pietro Bucci, Cubo 17/C, 87036 Rende (CS), Italy. E-mail: a.basile@itm.cnr.it; Phone: +39 0984 492013; Fax: +39 0984 402103.

In: Biochemistry Research Summaries. Volume 1
Editors: James T. Walker and Catherine M. Jones
ISBN: 978-1-53610-895-8

Chapter 129

ACETIC ACID IN MICROPALEONTOLOGY: ITS EFFICIENCY IN DISAGGREGATING CARBONATE ROCKS AND RECOVERING CALCAREOUS AND SILICEOUS MICROFOSSILS[#]

Marlone H. Hünnig Bom*[*], *Karlos Guilherme Diemer Kochhann, Gislaine Bertoglio Rodrigues and Gerson Fauth
Instituto de Micropaleontologia (ITT FOSSIL), Universidade do Vale do Rio dos Sinos, São Leopoldo, Rio Grande do Sul, Brazil

RESEARCH SUMMARY

Micropaleontology provides essential information for dating sedimentary rocks and characterizing depositional environments, mainly based on calcareous (*e.g.*, ostracods, foraminifers) and siliceous (*e.g.*, radiolarians) microfossils. Several methodologies are used for breaking down rocks and recovering microfossils. The composition, degree of lithification, porosity and organic matter content are features controlling rock disaggregation and microfossils recovery. Therefore, pelitic and carbonate rocks react differently to the application of different methods. There are several methodologies for extracting microfossils from carbonate rocks, which take into account chemical and/or physical properties of the microfossils to be recovered and the rock. These disaggregation techniques are revised herein considering the immersion of carbonate rocks in acetic acid and ranging kinetic parameters. Their efficiency in recovering both calcareous and siliceous microfossils, from samples that otherwise would be considered barren, is demonstrated. Besides, acetic acid methods are simple and easy to work, also permitting reutilization of residues for further reactions by means of distillation.

[#] This chapter was previously published in *Acetic Acids: Chemical Properties, Production and Applications*, edited by Angelo Basile, New York: Nova Publishers Inc., 2013.

[*] Corresponding author: marloneb@unisinos.br.

In: Biochemistry Research Summaries. Volume 1
Editors: James T. Walker and Catherine M. Jones
ISBN: 978-1-53610-895-8

Chapter 130

ACETIC ACID IN THE FOOD INDUSTRY[#]

Đurđica Ačkar*, Drago Šubarić,
Antun Jozinović and Jurislav Babić
Faculty of Food Technology Osijek, Osijek, Croatia

RESEARCH SUMMARY

Acetic acid has been known to humanity for many centuries. It was first discovered in the form of vinegar, when wine got spoiled by the action of bacteria. Ancient Romans cooked soured wine to produce highly sweet syrup.

Since then, the importance of acetic acid in food preparation/production steadily grew. Acetic acid used in food preparation/production can only be produced by natural process – aerobic fermentation, through action of acetic bacteria. In USA it is GRAS additive and in EU it is designated as E 260. It has no adverse effects and no dietary restrictions.

Today, acetic acid is not used only as vinegar in salad preparation. It is used in the preservation of food as a natural preservative and antimicrobial agent not only in vegetable pickling, but as an active component of edible films and in hurdle technology of preservation. It is added to mayonnaise to deactivate *Salmonella.* In addition, it is acidulant, gives a characteristic flavour profile to food, can be used for microbial decontamination of meat and as mild descaling agent in food industry. Research on acetic acid applications in the accelerated ripening of dates is also conducted.

Indole-3-acetic acid plays a role in banana ripening (phytohormone) and its degradation to 2-aminoacetophenon during wine production can give untypical aging off-flavour in wine. It can be used in grey mould control in harvested apple fruit.

Acetic acid and its anhydride are widely used in the production of modified starches, so called starch acetates as well as dual-modified starches.

This chapter will summarise current knowledge on acetic acid and its derivatives application in the food industry.

[#] This chapter was previously published in *Acetic Acids: Chemical Properties, Production and Applications*, edited by Angelo Basile, New York: Nova Publishers Inc., 2013.

* Corresponding author: Đurđica Ačkar, dackar@ptfos.hr.

In: Biochemistry Research Summaries. Volume 1
Editors: James T. Walker and Catherine M. Jones
ISBN: 978-1-53610-895-8

Chapter 131

COPRODUCTION OF VINEGAR AND ELECTRICITY BY APPLICATION OF MICROBIAL FUEL CELL TECHNOLOGY TO ACETIC ACID FERMENTATION[#]

Takanori Tanino*
Division of Environmental Engineering Science, Faculty of Science and Technology, Gunma University, Kiryu, Gunma, Japan

RESEARCH SUMMARY

It is well known that vinegar is one of the most familiar uses of acetic acid. In this chapter, the coproduction of vinegar and electricity by fermentation, which is a novel concept in vinegar production, is explained. The acetic acid in vinegar is produced from ethanol by acetic acid fermentation using acetic acid bacteria. Acetic acid fermentation is a typical example of a traditional application of microorganisms. In addition to traditional uses of microorganisms, the fields of microorganism application have been expanding owing to the progress of biotechnology. Among these fields, microbial fuel cell technology has recently been attracting attention as a novel method for harvesting energy from renewable organic compounds. The novel concept of the coproduction of vinegar and electricity has been developed by the application of microbial fuel cell technology to acetic acid fermentation. Acetic acid fermentation and microbial fuel cell technology share the same principle in that microorganisms are used as the catalyst for the oxidation of an organic compound. In this chapter, the core elements of acetic acid fermentation and microbial fuel cell technology that are important for understanding our novel concept are first described. Then, the advantages and our recent study of the coproduction of acetic acid and electricity are described.

[#] This chapter was previously published in *Acetic Acids: Chemical Properties, Production and Applications*, edited by Angelo Basile, New York: Nova Publishers Inc., 2013.

* Corresponding author: Takanori Tanino. Division of Environmental Engineering Science, Faculty of Science and Technology, Gunma University, 1-5-1, Tenjincho, Kiryu, Gunma 376-8515, Japan. E-mail: ttanino@gunma-u.ac.jp.

In: Biochemistry Research Summaries. Volume 1
Editors: James T. Walker and Catherine M. Jones
ISBN: 978-1-53610-895-8

Chapter 132

ORIGINS, PREVENTION AND TREATMENT OF CANCER#

***Salvatore de Salvo*[1] *and Mara Teresa de Salvo*[2]**
[1]Former Professor of Instituto Presbiteriano Mackenzie, Escola de Engenharia Mauá, Faculdade de Engenharia Industrial - FEI-PUC - SP, Faculdade de Engenharia da Fundação Armando Álvares Penteado, São Paulo, SP, Brazil
[2]MSc in Environmental Sanitation, former Professor at Faculdade de Engenharia da Fundação Armando Álvares Penteado, São Paulo, SP, Brazil

RESEARCH SUMMARY

To write about this disease, the most dangerous, ordinary and difficult to treat is, to say little, to dare to face a field of human activity so vast, complex and fascinating.

But before we specifically get into the issue, it is absolutely important to introduce the readers to the "DECLARATION OF HELSINKI," extracted from the World Medical Association, published in the "Journal da Associação Médica Brasileira" - AMB in June-July 2008. In order to spare the reader's time, instead of reproducing the entire declaration, we will reproduce only a few paragraphs, as for example from the introduction:

- Par. 2: It is a doctor's duty to promote and to safeguard the health of people. Both knowledge and awareness of doctors shall be dedicated to conducting such duty.
- Par. 4: Medical progress is based on research that in the last instance, must occur, often, in trials with humans.
- Par. 6: The primary purpose in research is to improve preventive procedures, diagnostic and therapeutic and the understanding of the etiology and pathogenesis of the disease. Even the best preventive methods, diagnostics and

This chapter was previously published in *Molecular and Supramolecular Bioinorganic Chemistry. Applications in Medical and Environmental Sciences. Volume 4*, edited by Ana Lucia Ramalho Mercê and María Ángeles Lobo Recio, New York: Nova Publishers Inc., 2013.

therapeutics, in a continuous manner, *shall* be challenged in the search for effectiveness, efficiency, quality and accessibility.

Among the basic principles for all medical research we want to highlight:

- Par. 11: Medical research involving humans has to be carried out according to generally accepted scientific principles, to be grounded on the complete knowledge of scientific literature, on other sources of relevant information and on adequate laboratory testing and when appropriate, on animal testing.
- Par. 18: Medical research involving humans should only be done if the purpose's importance exceed the inherent risks and charges for the subject. This is especially important when humans are healthy volunteers.

And among the applicable principles for medical research combined with medical care, we present:

- Par. 32: When, in the care of a patient, preventive, diagnostic and therapeutic methods result in proven ineffectiveness or is nonexistent, the physician, with the informed patient's consent, is allowed to use new or not yet proven preventive procedures, diagnostic or therapeutics, if, in his/her understanding, they will give any hope of saving a life, restoring health or relieving suffering. Whenever possible, such measures must be researched for the purpose of assessing the safety and effectiveness. In all cases, this new information must be recorded and, when appropriate, published. All other relevant policies of this statement must be observed.

Everything evolves in Nature. So does Medicine. Old concepts, old ideas also evolve, integrating what we call progress. And, in the history of mankind, there has never been an improvement as sharp, quick and stunning as in medical sciences. If we observe the rising curve of the recent scientific knowledge, we conclude that the advances in the last ten years exceed those achieved in the last 5,000 years... And if we continue this way, scientists claim that what will be discovered in the following years should become real not in 20 years but in 14, and in the following period, in only seven years.

This surprising and unexpected avalanche of new discoveries and inventions that are revolutionizing the world primarily in the communications field represents, in our view, the greatest demonstration that even though we humans inhabit such a tiny planet, absolutely insignificant in terms of the Galaxy and the universe, for some unknown reason we are and continue to be important. The intervention of superior intelligences to prepare us for the next quantum leap has never been so clear.

No wonder a deep knowledge of energy led to Nanotechnology, to a new Medicine, to new diagnostic methods, both preventive and therapeutic, alongside the discovery of ORMES (or ORMUS), among others. All this knowledge will allow this small planet to feed a population of up to 15 billion people, increasing the life span while ensuring that existential conditions are more acceptable. In fact, one cannot do anything without energy and while there are several alternative means of generating it nowadays, it should be noted that in the United States of America alone there are more than 4,500 patents shelved on how to extract and use the space-time free energy. This free energy, available anywhere and in infinite quantity, waves to a no longer remote but immediate possibility to recreate what we may consider the new terrestrial paradise.

What we will see below, briefly, will confirm what was said above.

In: Biochemistry Research Summaries. Volume 1
Editors: James T. Walker and Catherine M. Jones

ISBN: 978-1-53610-895-8

Chapter 133

NANOTECHNOLOGIES AND SUNSCREENS: NEW TECHNOLOGIES, IMPORTANCE AND RISKS[#]

Ana Cristina Trindade Cursino and Ana Lucia Ramalho Mercê*
Chemistry Department, Federal University of Paraná, Curitiba, PR, Brazil

RESEARCH SUMMARY

It is important to assess the effects of radiation on the human body considering individual characteristics, intensity, frequency and duration of sun exposure. The effects of radiation will also depend on the geographic location, season, time of day and weather conditions. A large surface of Brazil is located between the Tropic of Capricorn and the Equator. The sun's radiation in this region focuses at an angle closer to 90°, making Brazil the sunniest country with inter-tropical area of the planet. This fact, combined with the progression of the hole in the ozone layer can be a major contributor to the increased number of people presenting skin cancer in the country. To protect the skin from the harmful effects of ultraviolet radiation from the sun in order to reduce skin damage and the risk of skin cancer, it is recommended the use of physical photoprotective agents as clothing, hats, make-ups, sunglasses, window glasses, and ultraviolet light filters (sunscreen ingredients and sunless tanning agents).

[#] This chapter was previously published in *Molecular and Supramolecular Bioinorganic Chemistry. Applications in Medical and Environmental Sciences. Volume 4*, edited by Ana Lucia Ramalho Mercê and María Ángeles Lobo Recio, New York: Nova Publishers Inc., 2013.

[*] anamerce@ufpr.br.

In: Biochemistry Research Summaries. Volume 1
Editors: James T. Walker and Catherine M. Jones
ISBN: 978-1-53610-895-8

Chapter 134

INTERACTIONS OF ALUMINUM WITH ERYTHROID CELLS[#]

Daniela Vittori, Daiana Vota and Alcira Nesse*
Department of Biological Chemistry, School of Sciences,
University of Buenos Aires, Argentina
IQUIBICEN, National Council of Scientific and Technical Investigation,
Buenos Aires, República Argentina

RESEARCH SUMMARY

Aluminum (Al) is a widespread element used extensively in contemporary life, but despite its ubiquity, no essential role has been demonstrated for this metal in biological systems. It is not surprising, then, that this nonessential metal was long considered harmless to humans. The finding of a link between bone and neurological diseases in patients with renal failure under dialysis treatment and Al toxicity changed this belief. Nowadays, plenty of information suggests that, under certain environmental conditions, Al becomes toxic to plants, animals and humans. Further effort focused on this area will almost certainly improve our understanding of the influence of Al on biological systems.

[#] This chapter was previously published in *Molecular and Supramolecular Bioinorganic Chemistry. Applications in Medical and Environmental Sciences. Volume 4*, edited by Ana Lucia Ramalho Mercê and María Ángeles Lobo Recio, New York: Nova Publishers Inc., 2013.

* anesse@qb.fcen.uba.ar.

In: Biochemistry Research Summaries. Volume 1
Editors: James T. Walker and Catherine M. Jones
ISBN: 978-1-53610-895-8

Chapter 135

METAL-POLYSACCHARIDE COMPLEXES: AN IMPORTANT TOOL IN PHARMACOLOGICAL APPLICATIONS[#]

Ana Lucia Ramalho Mercê[1], Carmen Lúcia de Oliveira Petkowicz[2] and Guilhermina Rodrigues Noleto[2,*]

[1]Chemistry Department, Universidade Federal do Paraná, UFPR Curitiba, PR, Brazil
[2]Biochemistry Department, Universidade Federal do Paraná, UFPR Curitiba, PR, Brazil

RESEARCH SUMMARY

Metal-based drugs have had a major role as therapeutic agents in the ancient history of medicine and the pioneering times of modern pharmacology, especially as anti-infective, anti-cancer agents and anti-arthritic drugs (Reichert et al. 1999; Navarro et al. 2010). However, most of the metals frequently present high toxicity and severe side effects (Navarro et al. 2010). To reduce metallic ion toxicity, the concepts of covalent binding, chelation, coupling to proteins and amino-acids or carbohydrate polymers, inclusion into dextrins and liposomes have been investigated (Zhang and Lippard, 2003; Sibikina et al. 2009). In this chapter, it will be presented some applications of metal-polysaccharide complexes.

Carbohydrates function as structural or protective materials and as energy storage. In addition, carbohydrates are information-transfer agents. They are implicated in many important biological phenomena as in the process of infection by certain pathogens, determination of the human blood groups, cell-cell adhesion and cancer pathology (Dwek, 1996). The biological, chemical and physical properties of carbohydrates are determined by their structure and conformational behaviour. Differently from biological proteins and nucleic acids, carbohydrates can be highly branched molecules, and their monomeric units may be connected by many different linkage types (Dwek, 1996; Perez et al. 1996). This complexity

---

[#] This chapter was previously published in *Molecular and Supramolecular Bioinorganic Chemistry. Applications in Medical and Environmental Sciences. Volume 4*, edited by Ana Lucia Ramalho Mercê and María Ángeles Lobo Recio, New York: Nova Publishers Inc., 2013.

[*] guinoleto@yahoo.com.br.

allows carbohydrates to provide almost unlimited variations in their structures (Laine, 1994). Besides their importance in biological recognition systems, carbohydrates are the raw material for sustainable fuel, food, materials and medicine production (Persin et al. 2011).

Polysaccharides isolated from microorganisms (Konopski et al. 1994; Kulicke et al. 1997) as well as other sources have been recognized as biological response modifiers (BRM) (Bohn & BeMiller, 1995; Tzianabos, 2000; Schepetkins & Quin, 2006; Leung et al. 2006; Schepetkins et al. 2008; Zong et al. 2012). Besides native polysaccharides, derivatives arising from partial hydrolysis (Kulicke et al. 1997) or chemical modifications such as carboxymethylation (Bohn & BeMiller, 1995; Sasaki et al. 1979), oxidation and/or introduction of new groups, have also been successfully studied as BRM (Cross et al. 2001). Works in the literature showed remarkable results when this kind of compounds, extracted from different natural sources, can complex metallic ions providing very different behaviour than when not complexed (Mercê et al. 1998; 2001a,b; Mercê et al. 2002; Noleto et al. 2002; 2004; 2009). Interactions between carbohydrates and ions can display unique features in novel applications allowing potential uses in the pharmacological and pharmaceutical practice (Burger et al. 2001).

The interest of using the polysaccharides with metals in medicine and pharmaceutical industry has increased in the last decades. The aims are from diagnosis to therapeutic agents for different pathologies (Sibikina et al. 2009). In pharmacological applications, the carbohydrates or carbohydrate derivatives can act as metal carries (Webster, 1991, Cantos et al. 1993, Roberts et al. 1996; Caravan et al. 1995; Yuen et al. 1997, Matte et al. 2000; Crans et al. 2000; Sibikina et al. 2009; Oh et al. 2009; Dias et al. 2011) or concomitant administration with metal drugs to prevent side effects (Park et al. 2009; Liu et al. 2011; Zhang et al. 2002; Ohya et al. 2001).

A review published in 2009 collected many applications of carbohydrates, mainly in the presence of metal ions (Sibikina et al. 2009). Among them are the complexes with precious metals like platinum (cisplatin) and gold carbohydrate derivatives. There are also technological applications reports like wastewater remediations employing carbohydrates, (Sibikina et al., 2009 and references cited therein; Franco and Mercê, 2006; Franco et al. 2007; Hernández et al. 2007; Hernández et al. 2008, Lobo Recio et al. 2013), structural elucidation works (Mercê et al. 2002; Sibikina et al. 2009 and references cited therein) and medical applications of complexed carbohydrates (Sibikina et al. 2009 and references cited therein).

The supramolecular chemistry has also received a great contribution from the cyclodextrin capacity to host many compounds from pharmacological to industrial ones, to cite a few (Del Valle, 2004; Challa et al. 2005; Chaudhary and Patel, 2013). Cyclodextrins are reported in the literature not only to host molecules but also, as ternary complexes, to host metal complexes of some pharmaceutical drugs like Vitamin D_3 (Mercê et al. 2009, Soares et al. 2013).

Based on so many studies and importance of carbohydrates, this chapter will summarize some important aspects of this almost ubiquitous class of chemical substances and mainly when they are complexed to metal ions, which give them unique properties.

In: Biochemistry Research Summaries. Volume 1
Editors: James T. Walker and Catherine M. Jones
ISBN: 978-1-53610-895-8

Chapter 136

NOVEL TRENDS IN CYCLODEXTRINS: A REVIEW[#]

Edgar Winter Júnior, Daiane do Rocio Borguezani and Ana Lucia Ramalho Mercê*
Federal University of Paraná - Chemical Equilibrium Laboratory
LEQ – UFPR; Curitiba, PR, Brazil

RESEARCH SUMMARY

Cyclodextrins (CDs) have been known for nearly 100 years. CDs are naturally occurring water-soluble cyclic glucans. In 1881, Villiers first produced them by digesting starch with *Bacillus amylobacter* and in 1903, Schardinger demonstrated the cyclic structure of these compounds. They are produced as a result of intramolecular transglycosylation reaction from degradation of starch by cyclodextrin glucanotransferase enzyme (CGTase). CDs are also known as cycloamyloses, cyclomaltoses and Schardinger dextrins. The biologic reason for cyclodextrin synthesis is not clear. However, it is conceivable that some bacteria developed the producing and the hydrolyzing enzymes that allowed cyclodextrins to be used as a safe external storage medium.

CDs are a family of the macrocyclic oligosaccharides known as α-cyclodextrin (αCD), β-cyclodextrin (βCD), and γ-cyclodextrin (γCD), which are composed of six, seven, or eight α-(1,4) linked glycosyl units, respectively (Tables 1 and 2). Their hydrophobic cavity can be systematically varied and the hydroxyl groups, at the outer rims, can also be derivatized. Up to now, native CDs, α, β and γ CDs and their methylated, hydroxypropylated and carboxymethyl derivatives and pyrrolinone-modified β, γCD have been successfully used in many applications.

CDs are well-known as one of the most important host compounds in the Supramolecular Chemistry, so they have been extensively investigated because they act exactly as host molecules, or containers in the formation of inclusion compounds with many possible guest

---

[#] This chapter was previously published in *Molecular and Supramolecular Bioinorganic Chemistry. Applications in Medical and Environmental Sciences. Volume 4*, edited by Ana Lucia Ramalho Mercê and María Ángeles Lobo Recio, New York: Nova Publishers Inc., 2013.

* anamerce@ufpr.br.

molecules in aqueous solution. CDs can increase the aqueous solubility, chemical reactivity and spectral properties of numerous lipophilic drugs which are used as guest molecules, without changing their intrinsic ability to permeate the lipophilic membranes. Owing to their polar hydrophilic outer shells and hydrophobic cavities, one of the most important features of those compounds, is their capacity to host various substances (drugs, pesticides, perfumes and other products) in their hydrophobic inner cavities, thus forming encapsulated compounds. Each CD has its own ability to form host–guest inclusion complexes with specific guest molecules, an ability which depends on a proper fit of the guest molecule into the hydrophobic CD cavity.

In: Biochemistry Research Summaries. Volume 1
Editors: James T. Walker and Catherine M. Jones
ISBN: 978-1-53610-895-8

Chapter 137

ALUMINIUM, ADENOSINE-5'-TRIPHOSPHATE, PHOSPHOCREATINE AND AMINO ACIDS: HOW THEY CAN BE RELATED TO SOME NEURODEGENERATIVE DISEASES[#]

Thaís Tenório[1] and Ana Lucia Ramalho Mercê[2,*]
[1]Pontifícia Universidade Católica, Rio de Janeiro, RJ, Brazil
[2]Universidade Federal do Paraná, Curitiba, PR, Brazil

RESEARCH SUMMARY

There are numerous phosphate ligands in the body, e.g., adenosine-5'-triphosphate (ATP) and phosphocreatine (PCr), which present high levels in human plasma and in the brain, and also play critical roles in bioenergetics. Their chemical structure can be seen in Figure 1.

In humans, phosphorus is present in high amounts in tissues. In an adult of 70 kg body weight, 800 g is phosphorus, which amounts to approximately 1.14% of the total body mass. In biological systems, this element is in the form of inorganic (phosphate ions) or organic phosphates (phosphate esters) and it can participate in transfer reactions of phosphates, in the hydrolysis and in the phosphorolysis processes.

The absorption of phosphate occurs mainly in the inorganic state. The organically bound phosphate is hydrolyzed in the intestinal lumen by pancreatic and intestinal phosphatases, and so, inorganic phosphate is released, and subsequently, absorbed. The homeostasis and the excretion of phosphates are carried out by the kidneys. Excretion increases when there is greater ingestion or absorption of the element, in hyperparathyroidism, in metabolic respiratory acidosis, expansion of cell volume or by intake of diuretics.

[#] This chapter was previously published in *Molecular and Supramolecular Bioinorganic Chemistry. Applications in Medical and Environmental Sciences. Volume 4*, edited by Ana Lucia Ramalho Mercê and María Ángeles Lobo Recio, New York: Nova Publishers Inc., 2013.

* Email: anamerce@ufpr.br.

Phosphorus is an essential element that takes part in many bodily functions and metabolic processes, including: buffer action, which has a cooperation effect in the maintenance of the cytoplasmic nearly neutral pH, participates as the deoxyribonucleic acid (DNA) and ribonucleic acid (RNA) base, generates high-energy phosphate bonds in ATP, produces universal biological redutors like nicotinamide adenine dinucleotide (NADH) and nicotinamide adenine dinucleotide phosphate H (NADPH), constitutes the cyclic nucleotides, such as cyclic adenosine monophosphate (cAMP), which act as second messengers inside the cells and can activate kinases and phosphatases. In general, the functions of the phosphate compounds include organic structure formation, energy transfer and the initiation and regulation of cellular processes.

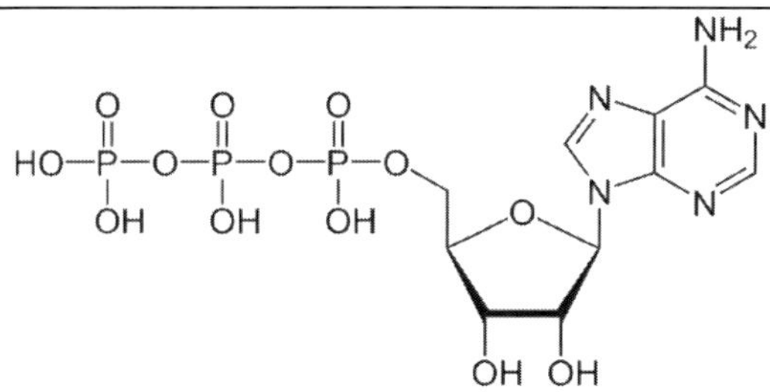
Adenosine-5'-triphosphoric acid [(2"R",3"S",4"R",5"R")-5-(6-aminopurin-9-yl)-3,4-dihydroxyoxolan-2-yl]methyl(hydroxyphosphonooxyphosphoryl)hydrogen phosphate Adenosine-5'-(tetrahydrogen triphosphate) Adenosine-5'-triphosphate
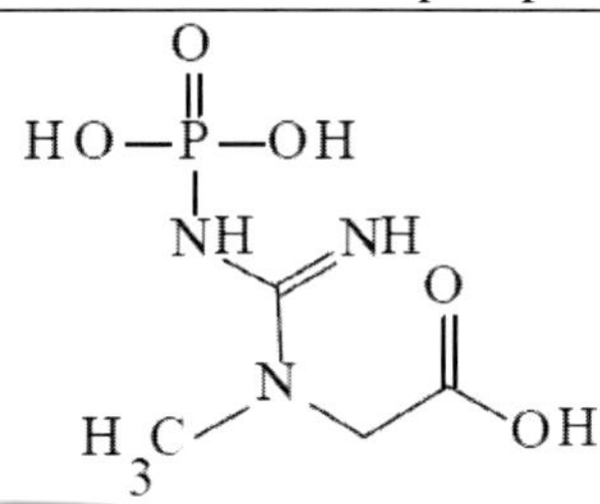
{[imino(phosphonoamino)methyl](methyl)amino}acetic acid N-Methyl-N-(N'-phosphonocarbamimidoyl)glycine Phosphocreatine

Figure 1. ATP and PCr structures and IUPAC nomenclature. (Adapted from http://en.wikipedia.org).

In: Biochemistry Research Summaries. Volume 1
Editors: James T. Walker and Catherine M. Jones
ISBN: 978-1-53610-895-8

Chapter 138

POLYAMINES, METAL IONS AND NEURODEGENERATIVE DISEASES: SOME CHEMICAL AND BIOLOGICAL ASPECTS[#]

Natalie Waissmann Szyfman[1], Judith Felcman[1,†] and Ana Lucia Ramalho Mercê[2,*]
[1]Pontifícia Universidade Católica, Rio de Janeiro, Brazil
[2]LEQ, Universidade Federal do Paraná, Curitiba, PR, Brazil

RESEARCH SUMMARY

Neurodegenerative diseases are characterized by apoptosis of cells in the central nervous system (CNS). Programmed cell death (PCD) is a highly regulated physiological process, which is essential and manages the cell turnover from fetal development to aging. During the development phase of the life of a multicellular organism, this process ensures a coordinated and clean disposal of cells that are no longer needed or that are inactive. Later, the process gradually takes over the functions of maintenance, the elimination of cells that are dysfunctional or damaged or otherwise hazardous to the body, together with its replacement. In rapidly proliferating tissues, namely the skin, gut, bone marrow and sexual organs of a human adult, approximately 60 billion cells are removed and replaced every day. Thus, cell death happens to everyone, in a regular process in the CNS. But, with advancing age, that loss of neurons accumulates and synapses, which are the nerve impulses which pass from one neuron to the next across the synaptic cleft through chemical mediators and neurotransmitters, responsible for stimuli, are harmed.

Thus neurodegenerative diseases, that are now epidemic, such as Alzheimer's disease (AD), Parkinson's disease (PD), frontotemporal lobar degeneration (FTLD), amyotrophic

---

[#] This chapter was previously published in *Molecular and Supramolecular Bioinorganic Chemistry. Applications in Medical and Environmental Sciences. Volume 4*, edited by Ana Lucia Ramalho Mercê and María Ángeles Lobo Recio, New York: Nova Publishers Inc., 2013.

[†] *In memoriam.*

[*] anamerce@ufpr.br.

lateral sclerosis (ALS), and vascular dementia (VD) are, according to some researchers, the result of greater longevity of people in the world. This is because people are reaching very advanced ages, and the accumulation of nerve cell loss becomes detrimental to them. And the prospect is that the elderly population will continue to grow.

These diseases are progressive and characterized by cognitive, motor and / or behavioral dysfunction. This clinical heterogeneity is largely attributed to the pathological variability and to the CNS region which is affected in every disease, with the latter being determined by the population of selective vulnerability of brain cells to the disease process.

The chains of globular proteins build structures with a stable conformation, but when these globular proteins undergo incorrect developments or undergo incorrect conformational change, as a result of a mutation of an amino acid, of chemical modification, of change in environment, or other unknown factors, proteins, often, are degraded or may aggregate and form amyloid plaques.

Factors such as abnormal aggregation of proteins and imbalance in the concentration of metal ions, represent a characteristic biochemical and histomorphologic of neurodegenerative disease. The classification of neurodegenerative diseases is increasingly based on the identity of the protein that accumulates. In most cases, these aggregates of proteins form intracellular deposits microscopically visible, the so-called "inclusion bodies". In most cases, these inclusion bodies are located in the cytoplasmic compartment. Thus, the study of the pathogenesis of the disease have historically focused on biochemical aspects of the anomalous protein that occurs in this part of the cell.

In: Biochemistry Research Summaries. Volume 1
Editors: James T. Walker and Catherine M. Jones
ISBN: 978-1-53610-895-8

Chapter 139

XENOBIOTICS IN THE ENVIRONMENT#

***Thaís Tenório*[1] *and Ana Lucia Ramalho Mercê*[2,*]**
[1]Pontifícia Universidade Católica do Rio de Janeiro - PUC - Rio
[2]LEQ - DQ - Universidade Federal do Paraná - UFPR - Brazil

RESEARCH SUMMARY

Xenobiotic agents are defined as any organic substance of low molecular weight (usually <1000 Da) foreign to the human organism, or which are not generated naturally in humans. They are chemical compounds, such as drugs, antibiotics and pollutants, natural and unnatural to which exposure is unavoidable. So, especially in a highly industrial society, mankind is exposed to daily contact with such compounds either through dietary habits, drugs or by pollutants.

Xenobiotics associated with eating habits can be food constituents, food additives and others such as coffee, tea, tobacco and alcohol. Furthermore, agrochemicals (fertilizers, insecticides, herbicides, pesticides) and industrial chemical products (dyes, solvents, polymers) are also xenobiotics, which may enter the human body. Such compounds when metabolized, are absorbed, distributed, biotransformed and excreted.

The organ responsible for the metabolism of foreign substances is the liver, but the gut and the lung can also participate in this process. The inflow of blood to the membrane of the hepatocytes occurs through active or passive transport depending on the type of xenobiotic to be metabolized. Most of lipophilic xenobiotics, for example, are transported by diffusion. The hepatic vectorial transport of xenobiotics is always mediated by transport and transfer proteins that are ligand-binding specific.

This chapter was previously published in *Molecular and Supramolecular Bioinorganic Chemistry. Applications in Medical and Environmental Sciences. Volume 4*, edited by Ana Lucia Ramalho Mercê and María Ángeles Lobo Recio, New York: Nova Publishers Inc., 2013.

* anamerce@ufpr.br.

In: Biochemistry Research Summaries. Volume 1
Editors: James T. Walker and Catherine M. Jones
ISBN: 978-1-53610-895-8

Chapter 140

THE ROLE OF CREATINE IN THE PATHOPHYSIOLOGY OF DEPRESSION AND THE POSSIBLE MECHANISMS UNDERLYING ITS ANTIDEPRESSANT EFFECT[#]

Mauricio P. Cunha, Francis L. Pazini, Ágatha Oliveira and Ana Lúcia S. Rodrigues
Department of Biochemistry, Center of Biological Sciences,
Universidade Federal de Santa Catarina, Florianópolis, SC, Brazil

RESEARCH SUMMARY

Depression is a common mood disorder associated with high rates of morbidity and mortality and characterized by a broad range of symptoms, including altered mood and cognitive functions, loss of interest or pleasure, feelings of guilt or low self-worth, disturbed sleep or appetite, low energy and recurrent thoughts of death or suicide. Although a number of new medications and therapies have been used for the management of this disorder, most of them were based on serendipitous discoveries. Consequently, the management of depression remains little changed over the last decades. However, accumulated evidence has suggested that an alteration in the brain energy metabolism (particularly a hypoactive prefrontal cerebral energy metabolism) may be implicated in the pathophysiology of this disease. It is therefore possible that agents that improve cellular bioenergetics may exert beneficial effects for treating depression. In this context, emerges creatine. The creatine kinase/phosphocreatine system plays a key role in the brain energy homeostasis by maintaining high ADP levels at the site in mitochondria at which ATP is generated and low levels at the site of ATP utilization. This chapter reviews the available findings that suggest that this compound exerts a role in the pathophysiology of depression and the possible mechanisms underlying its antidepressant effect.

[#] This chapter was previously published in *Creatine: Biosynthesis, Therapeutic Uses and Physiological Effects of Supplementation*, edited by Fernando D'Cruz and Victor Ribeiro, New York: Nova Publishers Inc., 2013.

In: Biochemistry Research Summaries. Volume 1
Editors: James T. Walker and Catherine M. Jones
ISBN: 978-1-53610-895-8

Chapter 141

THERAPEUTIC USES OF CREATINE: NEW POSSIBILITIES[#]

Martina Antošová[1] and Jana Plevkova[2]
[1]Department of Physiology,
[2]Department of Patological Physiology
Comenius University Jessenius Faculty of Medicine,
Martin, Slovakia

RESEARCH SUMMARY

Creatine is an organic acid that contribute to the energy supply to the skeletal muscle. The basic substrates for the creatine biosynthesis in the human body are semi-essential amino acids L-arginine, glycine and methionine. The creatine is synthesized in the liver and kidneys and partially can be substituted from food such as fish or meat. It is an essential source of muscle energy, which is used mainly by athletes to improve short-term performance. The creatine is one of the best-selling nutritional supplements. Effects of the creatine on increasing muscle strength was also tested in pathological conditions and diseases such as injuries or surgeries associated with prolonged immobilization, in patients with muscle weakness associated diseases, in patients with increased fatigue, in some neuromuscular diseases, in patients with muscular dystrophy, amyotrophic lateral sclerosis, chronic obstructive pulmonary disease or in patients with chronic heart failure. The creatine can be applied in various forms - creatine monohydrate, creatine phosphate, creatine maleate, creatine citrate etc. Based on available data we can conclude that the effect depends on the pharmacokinetic parameters of the forms of the creatine used for the supplementation. After per oral supplementation the creatine is rapidly and reversibly phosphorylated in the muscles. This reaction is catalyzed by creatine kinase, which produces creatine phosphate as a basic source of ATP resynthesis. ATP is subsequently used as a basic energy source. Use of the creatine is not introduced yet to standard practice, however some of the experimental and

---

[#] This chapter was previously published in *Creatine: Biosynthesis, Therapeutic Uses and Physiological Effects of Supplementation*, edited by Fernando D'Cruz and Victor Ribeiro, New York: Nova Publishers Inc., 2013.

clinical studies indicate that supplementation of the creatine improves function of skeletal muscle in some diseases (such as COPD), and consequently the quality of life of affected patients. The creatine supplementation can replace anabolic steroids or nutritional supplements in conditions associated with malnutrition, such as cancers or COPD. Very interesting areas represent neuroprotective effects of creatine, effects on cognitive and mental functions, influence on the metabolism of carbohydrates and fats. Currently, the use of the creatine is very popular especially among young people and athletes in order to increase muscle strength, muscle growth and weight loss. The creatine may also, as mentioned above, improve the performance especially in relation to short-term sports activity. The results of the creatine supplementation in the pathological conditions remain controversial. The studies reported positive results but the effects of the creatine often depend on many factors like comorbidities and individual metabolic differences. The review summarizes the current therapeutic possibilities of the creatine as well as the results of recent research activity in the creatine field.

In: Biochemistry Research Summaries. Volume 1
Editors: James T. Walker and Catherine M. Jones
ISBN: 978-1-53610-895-8

Chapter 142

MUSCLE ERGOGENIC EFFECTS OF CREATINE SUPPLEMENTATION IN RESISTANCE EXERCISE TRAINING[#]

Roberto Carlos Burini[1,*] and Edilson Serpeloni Cyrino[2]
[1]Centre for Nutrition and Exercise Metabolism, Botucatu Medical School, Sao Paulo State University (UNESP), Botucatu (SP), Brazil
[2]Center of Physical Education and Sports, Londrina State University Londrina (PR), Brazil

RESEARCH SUMMARY

Creatine in its monohydrated form (CrM) is on the top of the list among athletes seeking ergogenic supplements for increasing muscle mass and/or strength. It is proposed that CrM supplementation induces cell swelling which would activate downstream cell-volume signaling cascades and affect muscle-cell metabolism. The up regulated protein involved in osmosensing would result in increases in maximal muscle strength and power, fat-free mass, total body water and body weight. These would happen independently of training but overall CrM supplementation actions usually potentiate concurrent heavy-resistance training effects on muscle. CrM has also a major role on post-exercise induced damage by slowing down the degeneration phase and accelerating the regeneration phase after high-intensive exercises. As such CrM may provide protective effects via increased phosphocreatine synthesis recovering the ATP levels and maintaining the Ca^{2+}-ATPase pump. Such effects would keep the Ca^{2+} inside the sarcoplasmatic reticulum avoiding Ca^{2+} dependent proteases activation, protein breakdown and higher cell membrane permeability (intracellular components leakage). Traditional CrM supplementation protocols include phases of loading and maintenance. Co-ingestion of CrM with carbohydrate and protein may enhance Cr uptake through insulin action. Although there may be various side effects of Cr supplementation the only clinically

---

[#] This chapter was previously published in *Creatine: Biosynthesis, Therapeutic Uses and Physiological Effects of Supplementation*, edited by Fernando D'Cruz and Victor Ribeiro, New York: Nova Publishers Inc., 2013.
[*] Corresponding author: Email: burini@fmb.unesp.br.

significant side effect scientifically proven has been body weight gain. Nevertheless, it is recommended to avoid high dosages of Cr during periods of increased thermal stress as those under high ambient temperature/humidity. Thus Cr supplementation can be considered a safe, ethical, legal, inexpensive and effective nutritional intervention particularly when consumed in conjunction with a resistance training protocol.

In: Biochemistry Research Summaries. Volume 1
Editors: James T. Walker and Catherine M. Jones
ISBN: 978-1-53610-895-8

Chapter 143

EXPERIMENTAL EVIDENCE THAT CREATINE SUPPLEMENTATION DURING PREGNANCY IS PROTECTIVE FOR THE NEONATE[#]

David W. Walker PhD, DSc,[1,2,*] Hayley Dickinson PhD,[1] Stacey Ellery,[1] Domenic LaRosa,[1] Zoe Ireland PhD,[3] Syed Baharom,[1,4] Richard Harding PhD, D.Sc,[4] and Rod Snow PhD[5]

[1]The Ritchie Centre, Monash Institute for Medical Research, Clayton, Melbourne, Australia
[2]Department of Obstetrics and Gynaecology, Monash Medical Centre, Monash University, Clayton, Victoria, Australia
[3]Queensland Cerebral Palsy and Rehabilitation Research Centre, Royal Brisbane and Women's Hospital, Herston, Queensland, Australia
[4]Department of Anatomy and Developmental Biology, Monash University, Clayton, Victoria, Australia
[5]Centre for Physical Activity and Nutrition Research, School of Exercise and Nutrition Sciences, Deakin University, Burwood, Victoria, Australia

RESEARCH SUMMARY

The ergogenic and neuroprotective properties of creatine are now well recognised, and creatine supplementation is used widely by healthy adults and as a therapy for patients with neurological and musculo-skeletal conditions where depletion of ATP might be prevented by increasing the intracellular creatine pool. Babies can be born under conditions where poor

---

[#] This chapter was previously published in *Creatine: Biosynthesis, Therapeutic Uses and Physiological Effects of Supplementation*, edited by Fernando D'Cruz and Victor Ribeiro, New York: Nova Publishers Inc., 2013.

[*] Corresponding author: David W. Walker, The Ritchie Centre, Monash Institute of Medical Research, 27-31 Wright St., Clayton, Melbourne, Australia, 3168. Tel.: 613-9594 5372; Fax: 613 – 9594 6249.

placental perfusion, intrauterine infection, or depressed respiratory drive result in oxygen deprivation, increasing the risk of damage arising in the brain and other key organs of the neonate due to the inability of cells to rapidly replenish ATP via the creatine kinase pathway. This chapter reviews recent experimental work in pregnant and newborn animals that shows the benefits of creatine supplementation for survival after birth when oxygen and nutrient supply may limit cellular respiration, providing the basis for further consideration of the use of creatine in human pregnancy.

In: Biochemistry Research Summaries. Volume 1
Editors: James T. Walker and Catherine M. Jones
ISBN: 978-1-53610-895-8

Chapter 144

CREATINE: METABOLISM AND ROLE IN SPORTS PHYSIOLOGY[#]

Ayed Ikram Bezrati-Ben[1], Fehmi Nasrallah[2], Raouf Hammami[3], Karim Chamari[4], Moncef Feki[5] and Naziha Kaabachi[5*]

[1]Laboratory of Biochemistry, Rabta Hospital, Tunisia;
National Centers for Medicine and Sciences of Sports,
Faculty of Medicine of Tunis, El Manar University, Tunis, Tunisia
[2]Laboratory of Biochemistry, Rabta Hospital, Tunisia
[3]Research Laboratory "Sport Performance Optimization"
National Centers for Medicine and Sciences of Sports, Tunisia
[4]Research & Education Center Aspetar, Doha, Qatar
[5]Laboratory of Biochemistry, Rabta Hospital, Tunisia;
Faculty of Medicine of Tunis, El Manar University, Tunis, Tunisia

RESEARCH SUMMARY

Methyl guanidine-acetic acid, commonly named creatine (Cr), is a naturally occurring amino acid provided by diet or synthesized endogenously, primarily in liver, pancreas and kidneys. Cr biosynthesis involves two enzymes; arginine glycine amidinotransferase that converts arginine and glycine to ornithine and guanidinoacetate, and S-adenosyl-L-methionine:N-guanidinoacetate methyltransferase that converts guanidinoacetate to Cr. Cr is taken up from blood into Cr-requiring tissues against a concentration gradient by a specific transporter (SLC6A8). Approximately 95% of Cr is stored in skeletal muscle and the remaining 5% is stored in the heart, brain and testes. Cr and its derivative phosphocreatine (PCr) play an essential role in energy storage and transmission in most tissues; predominant

[#] This chapter was previously published in *Creatine: Biosynthesis, Therapeutic Uses and Physiological Effects of Supplementation*, edited by Fernando D'Cruz and Victor Ribeiro, New York: Nova Publishers Inc., 2013.

* Corresponding author: Naziha Kaabachi, Laboratory of Biochemistry, Rabta Hospital, 1007 Jebbari, Tunis, Tunisia. Tel/fax, 00 216 71 561 912; E-mail address: Naziha.kaabachi@rns.tn; Naziha.kaabachi@gmail.com

role being played in skeletal muscle and brain. In athletes, PCr is particularly important during explosive activities when a high rate of energy release is required. PCr serves as a source of high energy phosphates and increasing PCr muscle stores might have a significant effect on physical performance, especially during anaerobic intense exercises. Several researchers have evaluated the potential ergrogenic value of Cr. It has been shown that Cr supplementation increases total Cr content in skeletal muscle, particularly in individuals with lowest basal Cr content. Numerous studies have examined the effect of Cr supplementation in athletes. Most studies have shown that Cr supplementation is accompanied by a significant increase in body weight, with increase of lean mass and decrease of fat mass. This may be due to stimulation of muscle protein synthesis or water retention, a point that is particularly relevant for athletes in explosive sports disciplines. The results support the proposal that Cr supplementation improves performance particularly during high intensity exercise, as well as resistance training. However, other studies failed to show a significant effect of Cr supplementation. Issues such as the dose and duration of supplementation, exercise protocol, and sports disciplines may have contributed to these discrepancies. There are some anecdotal reports on potentially adverse effects of Cr supplementation such as muscle cramps, gastrointestinal complaints, liver dysfunction, and kidney impairment. It appears that Cr supplementation at the correct dose is not harmful to healthy athletes, but it must be used with precaution in individuals at risk. Finally, it seems reasonable to monitor liver and kidney functions in individuals consuming Cr supplements.

In: Biochemistry Research Summaries. Volume 1
Editors: James T. Walker and Catherine M. Jones

ISBN: 978-1-53610-895-8

Chapter 145

GLYCOGEN STORAGE DISEASE TYPE II: EXPANDING THE COMPREHENSION OF CLINICAL AND GENETIC SPECTRUMS OF A GENERALIZED DISEASE[#]

Gauthier Remiche[1,2,*], Maurizio Moggio[3] and Giacomo P. Comi[4]

[1]Fonds Erasme pour la Recherche Médicale Université Libre de Bruxelles (ULB), Bruxelles, Belgium

[2]Department of Neurology, Hôpital Erasme, Université Libre de Bruxelles (ULB), Brussels, Belgium

[3]Neuromuscular and Rare Diseases Unit, IRCCS Foundation Ca' Granda Ospedale Maggiore Policlinico, Milan, Dino Ferrari Centre, University of Milan, Italy

[4]Dino Ferrari Centre, Neuroscience Section, Department of Pathophysiology and Transplantation (DEPT), University of Milan, Neurology Unit, IRCCS Foundation Ca' Granda Ospedale Maggiore Policlinico, Milan, Italy

RESEARCH SUMMARY

Glycogen Storage Disease Type II (GSDII) is an autosomal recessive lysosomal storage disease due to alpha acid glucosidase (GAA) deficiency (*GAA* gene, 17q25.2-q25.3), a ubiquitary lysosomal enzyme that degrades glycogen into glucose. GSDII prevalence varies from 1:40000 to 1:146000. Most authors consider only the infantile and late onset form, the latter being a large continuum.

More than 350 pathogenic mutations have been found and some of them concern the majority of patients.

[#] This chapter was previously published in *Glycogen: Structure, Functions in the Body and Role in Disease*, edited by Pedro L. Weiss and Brian D. Faulkner, New York: Nova Publishers Inc., 2013.

[*] Corresponding author: Gauthier Remiche, MD. Neurologie, ULB Erasme, 808 Route de Lennik, 1070 Bruxelles, Belgium. Phone: 32 2 555 39 92. Fax: 32 2 555 39 42. E-mail: gauthier.remiche@erasme.ulb.ac.be.

In the neonatal period GAA mobilizes glycogen from the liver and later it avoids its accumulation in the lysosomes but its exact role(s) in adulthood remains uncertain.

Several stages of glycogen accumulation have been described and autophagia probably plays a central role in the process leading to muscle tissue damage. Glycogen accumulates in numerous organs including skeletal muscle, heart, liver, smooth muscle and the anterior horn of the spinal cord.

In the infantile form, the clinical picture is a floppy infant with severe respiratory distress, heart and liver failure. Death commonly occurs before one year of life. In some cases the first symptoms can occur later but by definition, before two years of life. While the disease generally leads to severe disability and premature death due to respiratory failure, the late onset form represents a spectrum of phenotypes from the mildest to the most severe.

Compared to other genetically-determined myopathies, GSDII causes a disproportionate respiratory failure mainly due to diaphragm weakness. Other particular phenotypical features were described as tongue enlargement, ptosis, hypoacusia, arterial abnormalities, incontinence or hydromyelia.

No consensus is currently admitted concerning the possible complex relationship between genotype and phenotype.

The diagnosis, in late onset form, can be challenging as ancillary tests can be uninformative. It is based on a low level of GAA activity and can be confirmed by the demonstration of biallelic pathogenic mutation.

Muscle biopsy can show typical abnormalities such as vacuolar myopathy with PAS positive and/or highly reactive acid phosphatase vacuoles.

The management of patients includes physiotherapy, the detection and care of respiratory failure. Enzyme replacement therapy (ERT) has shown its efficiency concerning disability progression and life expectancy but the response to the treatment varies among the patients.

In conclusion, GSDII is a rare multisystemic lysosomal storage disease that displays a wide spectrum of phenotypes and prognoses. The recent progresses of diagnosis and follow-up skills, and, the new care approaches including ERT are promising to improve the prognosis of patients.

In: Biochemistry Research Summaries. Volume 1
Editors: James T. Walker and Catherine M. Jones
ISBN: 978-1-53610-895-8

Chapter 146

GLYCOGEN METABOLISM DURING CHRONIC LIVER DISEASES[#]

***Natalia N. Bezborodkina*[*], *Anna Yu. Chestnova*[†] and *Boris N. Kudryavtsev*[‡]**
Laboratory of Cellular Pathology, Institute of Cytology of the Russian Academy of Sciences, Saint Petersburg, Russia

RESEARCH SUMMARY

The chapter focuses on the metabolism of glycogen and glucose during chronic hepatitides and cirrhosis of various etiology. Chronic hepatitides and their final stage, liver cirrhosis, are common and severe human diseases. They cause the impairment of the liver's morphological structure as well as various changes in the metabolism of carbohydrates, including that of glycogen, which is reflected in the glucostatic function of the liver. In particular, carbohydrate metabolism during cirrhosis is similar to that during diabetes, whose main feature is tissue intolerance to glucose. The chapter mostly presents the results of the authors' original research. Carbohydrate metabolism in the normal and the pathologically changed liver is examined during starvation and at various stages of the trophic cycle. The chapter contains a detailed discussion of regulatory mechanisms of glycogen metabolism in the normal and the cirrhotic liver, which help maintain the glucose homeostasis during starvation, in the absorptive and post-absorptive periods of the trophic cycle. Data on the activity of several key enzymes of glycolysis, gluconeogenesis, glycogenesis and glycogenolysis are given. Much attention is given to the glucose and glycogen metabolism under conditions of changing severity of the liver's chronic impairment.

[#] This chapter was previously published in *Glycogen: Structure, Functions in the Body and Role in Disease*, edited by Pedro L. Weiss and Brian D. Faulkner, New York: Nova Publishers Inc., 2013.

[*] Corresponding author N. N. Bezborodkina, Ph.D. Laboratory of Cellular Pathology, Institute of Cytology of the Russian Academy of Sciences, Tikhoretsky ave 4, 194064 Saint Petersburg, Russia. E-mail: natalia_bezborodkina@mail.ru.

[†] E-mail: chestnovaanna@mail.ru

[‡] E-mail: bn_kudryavtsev@mail.ru.

In: Biochemistry Research Summaries. Volume 1
Editors: James T. Walker and Catherine M. Jones

ISBN: 978-1-53610-895-8

Chapter 147

MUSCLE AND LIVER GLYCOGEN IN HELMINTH-INFECTED FISH[#]

Galina I. Izvekova [*] ***and Evgeny I. Izvekov***
Papanin Institute for Biology of Inland Waters, Russian Academy of Sciences, Borok, Nekouzskii Raion, Yaroslavskaya Oblast, Russia

RESEARCH SUMMARY

The influence of trematode and cestode infections on the glycogen content in the muscles and liver of various fish species was studied. The glycogen content was found to increase in the muscles of young cyprinid fishes (ide *Leuciscus idus*, dace *Leuciscus leuciscus*, goldfish *Carassius auratus*, and roach *Rutilus rutilus*) following the experimental infection with the metacercariae of the trematode *Apophallus muehlingi*. The same was true for the roach juveniles naturally infected with another trematode *Posthodiplostomum cuticola*. These changes could be explained by a decrease in swimming activity of fish due to parasite-induced muscle injuries. As an example, a significant drop in spontaneous motor activity was shown for the crucian carp *Carassius carassius* after 30-day maintenance with the trematode-shedding mollusks, which may be a possible reason for glycogen accumulation in the muscles of infected fish.

At the same time, glycogen concentrations in the muscles and liver declined when the fish (perch *Perca fluviatilis*, bream *Abramis brama*, and sabrefish *Pelecus cultratus*) were infected with the plerocercoids of some cestodes (*Triaenophorus nodulosus* and *Ligula intestinalis*). Reduced glycogen content in the liver testifies to a dysfunction of this organ in the infected fish and, as a consequence, leads to a decrease in the activity of major digestive enzymes (proteinases and glycosidases) in the intestine. Thus, the infection with trematodes and cestodes differently affects the glycogen content in the fish tissues and, respectively, might cause different consequences for animal health.

[#] This chapter was previously published in *Glycogen: Structure, Functions in the Body and Role in Disease*, edited by Pedro L. Weiss and Brian D. Faulkner, New York: Nova Publishers Inc., 2013.

[*] Corresponding author: izvekov@ibiw.yaroslavl.ru.

In: Biochemistry Research Summaries. Volume 1
Editors: James T. Walker and Catherine M. Jones
ISBN: 978-1-53610-895-8

Chapter 148

GLYCOGEN METABOLISM ENZYMES AS MOLECULAR TARGETS FOR DRUG DEVELOPMENT[#]

Vassiliki T. Skamnaki, Anastassia L. Kantsadi, Demetra S. M. Chatzileontiadou, Georgios Stravodimos and Demetres D. Leonidas*
Department of Biochemistry and Biotechnology, University of Thessaly, Greece

RESEARCH SUMMARY

Glycogen, the cellular energy store, is associated with a set of enzymes controlling its metabolism in a complex and dynamic way. Although glycogen metabolism has been thorough investigated, the regulatory mechanisms governing this process are not fully understood to date. New ideas have emerged in an effort to elucidate the complications of regulation of glycogen metabolism, from altered phosphorylation status and enzyme translocation to the role of intracellular compartmentalization and heterogeneity of glycogen stores as a regulator of metabolic pathways, and signalling transduction and lysosomal disposal of glycogen. The correlation of glycogen metabolism to many metabolic disorders has targeted many of the enzymes involved, for the development of drugs for a number of metabolic diseases, predominantly diabetes mellitus but also cancer. The notion of rational drug design emerged as the most appropriate modern interdisciplinary approach. This approach involves the close collaboration of diverse fields from bioinformatics and organic synthesis to structural biology (X-ray crystallography, NMR and EM) and physiology. In this review we will attempt to highlight major aspects and recent progress in the research for new drugs targeting enzymes playing a dominant role in the degradation rather than the synthetic glycogen pathway, such as glycogen phosphorylase and glycogen phosphorylase kinase.

[#] This chapter was previously published in *Glycogen: Structure, Functions in the Body and Role in Disease*, edited by Pedro L. Weiss and Brian D. Faulkner, New York: Nova Publishers Inc., 2013.

* To whom correspondence should be addressed: Assoc. Prof. Demetres D. Leonidas, Department of Biochemistry and Biotechnology, University of Thessaly, 26 Ploutonos Str. 41221 Larissa, Greece. Phone: +30 2410 565278; Fax: +30 2410 565290;e-mail: ddleonidas@bio.uth.gr.

In: Biochemistry Research Summaries. Volume 1
Editors: James T. Walker and Catherine M. Jones
ISBN: 978-1-53610-895-8

Chapter 149

THE LINK BETWEEN OMEGA-3 FATTY ACIDS AND RHEUMATOID ARTHRITIS: PROPERTIES, MECHANISMS AND THERAPEUTIC EFFICACY[#]

Gerard G. Dumancas[1,*], Rangika S. Hikkaduwa Koralege[2], Elmer-Rico E. Mojica[3], Befrika S. Murdianti[4], and Patrisha J. Pham-Bugayong[5]

[1]Arthritis and Clinical Immunology Research Program, MS, Oklahoma Medical Research Foundation, Oklahoma City, OK, US
[2]Department of Chemistry, Physical Science, Oklahoma State University, Stillwater, OK, US
[3]Department of Chemistry and Physical Sciences, Pace University, One Place Plaza, New York, NY, US
[4]Department of Physical Sciences, Arkansas Tech University, Russellville, AR, US
[5]Dave C. Swalm School of Chemical Engineering, Mississippi State University, MS, US

RESEARCH SUMMARY

Omega-3 fatty acids (FAs) are long chain polyunsaturated fatty acids characterized by a double (C=C) bond starting after the third carbon atom from the end of the carbon chain. These FAs are known to be primarily abundant in fish and plant oil products. Numerous studies indicate that omega-3 FAs have anti-inflammatory properties and, therefore, might be useful in the management of inflammatory and autoimmune diseases such as Rheumatoid Arthritis (RA). Rheumatoid arthritis (RA) is one of the systemic inflammatory disorders that gained widespread importance in every aspect of research due to its controversy of reducing

---

[#] This chapter was previously published in *Omega-3 Fatty Acids: Chemistry, Dietary Sources and Health Effects*, edited by Md. Wasim Khan, New York: Nova Publishers Inc., 2013.

[*] Corresponding author: Gerard G. Dumancas, Arthritis and Clinical Immunology Research Program, MS58, Oklahoma Medical Research Foundation, 825 N.E. 13th Street, Oklahoma City, OK 73104. E-mail: gerard-dumancas@omrf.org.

the lifespan of patients from 3 to 12 years. It is a common disorder that affects both men and women at the prime of their lives with extra-articular manifestations that include vasculitis, lung disease, pericarditis, neuropathy, and subcutaneous nodules. This chapter will discuss the pharmacological properties, mechanisms, and therapeutic efficacy of various omega-3 FAs as it relates to RA, as well as the importance of omega-6/omega-3 ratio balance to such disease.

In: Biochemistry Research Summaries. Volume 1
Editors: James T. Walker and Catherine M. Jones

ISBN: 978-1-53610-895-8

Chapter 150

OMEGA-3 FATTY ACIDS TARGET NF-κB TO MODULATE BREAST AND PROSTATE CANCER PROGRESSION[#]

D. A. Cavazos, S. A. Apte, C. H. Chen and L. A. deGraffenried
Department of Nutritional Sciences, University of Texas at Austin, Austin, Texas, US

RESEARCH SUMMARY

Background: Global epidemiological trends suggest that consumption of dietary sources enriched in omega-3 fatty acids correlate with improved clinical benefit in the prevention and treatment of certain cancers. In this chapter, we will focus primarily on the chemopreventive and chemotherapeutic properties of cold-water fish-derived omega-3 fatty acids. Numerous studies have found that omega-3 fatty acids suppress the formation and growth of a variety of solid tumors, including carcinomas of the breast and prostate. Here we will review the current literature and present our *in vitro* mechanistic findings which support the role of omega-3 fatty acids in decreasing risk of breast and prostate cancer development through modulation of nuclear factor kappa-light-chain-enhancer of activated B cells (NF-κB).

Key Findings: NF-κB is a transcription factor involved in regulating expression of pro-inflammatory and pro-survival genes in response to various forms of cellular stress, and is commonly dysregulated in solid tumors. Previous studies by our group and others provide evidence that docosahexaenoic acid (DHA), an omega-3 fatty acid, suppresses NF-κB transcriptional activity and precludes translocation into the nucleus in two human prostate tumor-derived cell lines, LNCaP and PacMetUT1. Additionally, inhibition of NF-κB was associated with several parameters of physiological response, including decreased proliferation and survival. In addition to the beneficial effects of DHA in the prostate, we observe complimentary and potent chemotherapeutic effects in breast cancer tumor cell lines. MCF-7 (ER+) and MDA-MB-231 (ER-) cells treated with omega-3 fatty acids also display nuclear exclusion of NF-κB and reduced proliferation and survival. Together these findings

[#] This chapter was previously published in *Omega-3 Fatty Acids: Chemistry, Dietary Sources and Health Effects*, edited by Md. Wasim Khan, New York: Nova Publishers Inc., 2013.

suggest that dietary omega-3 fatty acids elicit antiproliferative and pro-apoptotic effects in prostate and breast tumor cell lines.

Conclusions: Preclinical and epidemiologic studies indicate that omega-3 fatty acids improve clinical outcome in prostate and breast cancers. We provide *in vitro* evidence that omega-3 fatty acids function as a chemotherapeutic agent in breast and prostate carcinomas, in part through inhibition of NF-κB.

In: Biochemistry Research Summaries. Volume 1
Editors: James T. Walker and Catherine M. Jones
ISBN: 978-1-53610-895-8

Chapter 151

OMEGA-3 FATTY ACIDS AND CARDIOVASCULAR DISEASES[#]

***Federico Guerra*[*], Luca Piangerelli, Andrea Romandini, Simone Maffei and Alessandro Capucci**
Cardiology and Arrhythmology Clinic, Center for the Study of Obesity, Marche Polytechnic University, Italy

RESEARCH SUMMARY

Since the expeditions to Greenland by Bang and Dyerberg starting in the late 1960s, a diet rich of omega-3 fatty acids was suggested as the main responsible for the very low incidence of myocardial infarction in the Inuit population when compared to Danish controls. A few decades after, omega-3 fatty acids have been reported as associated with an antiatherogenic blood lipid pattern, reduced platelet reactivity, and fewer cardiovascular events. Proposed mechanisms for the protective role of omega 3 fats against cardiovascular diseases include blood pressure lowering; altered lipid profile, especially reduced serum triglyceride concentration; reduced thrombotic tendency; anti-inflammatory effects; anti-arrhythmic effects including heart rate reduction; improved vascular endothelial function; increased plaque stability; increased paraoxonase levels and improved insulin sensitivity. Unfortunately, while previous randomized clinical trials presented some enthusiastic results regarding a strong positive role of omega-3 fatty acids in preventing death, myocardial infarction, stroke and ventricular arrhythmias, recent ones failed to confirm such benefits on broader populations. Moreover, different dosages of either eicosapentaenoic acid (EPA), docosahexaenoic acid (DHA) or both, administered either through dietary counseling or tablet supplement, were used while collecting evidences. Systematic reviews and meta-analyses

---

[#] This chapter was previously published in *Omega-3 Fatty Acids: Chemistry, Dietary Sources and Health Effects*, edited by Md. Wasim Khan, New York: Nova Publishers Inc., 2013.

[*] Cardiology and Arrhythmology Clinic, Center for the Study of Obesity, Marche Polytechnic University; Email: f.guerra@univpm.it.

added further confusion to the existing controversy as they often reported conflicting findings.

Recent guidelines issued by major cardiology societies broadly recommend their use, either as supplements or through dietary counseling, but for different settings. While is widely accepted the use of omega-3 fatty acids for secondary prevention in patients with a recent myocardial infarction, some European countries have approved the omega-3 supplements for lowering cardiovascular risk whereas the US Food and Drug Administration has approved their administration only as triglyceride-lowering agents in patients with overt hypertriglyceridemia. In the present chapter the authors will summarize currently available evidence in order to clarify the role of omega-3 fatty acids in the cardiovascular setting and define the best spot for this new therapeutics.

In: Biochemistry Research Summaries. Volume 1
Editors: James T. Walker and Catherine M. Jones
ISBN: 978-1-53610-895-8

Chapter 152

FISH OIL IN COLON CANCER PREVENTION: CAN BIOMARKERS TELL US HOW MUCH IS ENOUGH?[#]

Yan Jiang[1], Dean E. Brenner[1,2,3], William L. Smith[4] and Zora Djuric[5*]

[1]Department of Internal Medicine,
[2]Department of Pharmacology, University of Michigan Medical School,
University of Michigan, US
[3]VA Medical Center, Ann Arbor, MI, US
[4]Department of Biological Chemistry,
[5]Department of Family Medicine
University of Michigan Medical School,
University of Michigan, Ann Arbor, MI, US

RESEARCH SUMMARY

Experimental studies, performed *in vivo* and *in vitro*, have provided strong indications for the anti-tumor actions of omega (ω) 3 fatty acids from fish oil against colon cancer. The molecular mechanisms of action have also been investigated extensively. However, the heterogeneity of the models utilized in these studies do not allow for prediction of the optimal dose of fish oil for healthy individuals. Many nutrients or supplements have non-linear dose-response curves with regard to beneficial effects. Whether or not higher doses of ω-3 fatty acids will increase the efficacy of fish oil in terms of reducing the risk of colon cancer is still an unanswered question. In this chapter, the available evidence for the beneficial effects of ω-3 fatty acids from fish oil supplements for colon cancer prevention is presented with a view towards identifying possible minimum and maximum fish oil intakes that would lead to a

[#] This chapter was previously published in *Omega-3 Fatty Acids: Chemistry, Dietary Sources and Health Effects*, edited by Md. Wasim Khan, New York: Nova Publishers Inc., 2013.

* Corresponding Author: Zora Djuric, Ph.D. Email: zoralong@med.umich.edu.

preventative effect. It is also suggested that biomarkers, such as the ω-3/ω-6 fatty acid ratio, may serve to personalize the efficacy of fish oil supplementation for reducing the risk of colon cancer.

In: Biochemistry Research Summaries. Volume 1 ISBN: 978-1-53610-895-8
Editors: James T. Walker and Catherine M. Jones

Chapter 153

EPIDERMAL BARRIER FUNCTION AND OMEGA-3 FATTY ACID SUPPLEMENTATION

***Audrey Lajoinie*[1,2], *Patrick Gelas*[3], *Marek Haftek*[1] and *Fabrice Pirot*[1,2]**

[1]University Lyon 1 – Claude Bernard, "Physiological, Clinical and Therapeutic Aspects of the Skin Barrier Function", Lyon, France
[2]Hospices Civils de Lyon –Edouard Herriot Hospital, Pharmacy, Plateform FRIPHARM, Lyon, France
[3]Hospices Civils de Lyon, Croix-Rousse Hospital, Clinical Nutrition Intensive Care Unit, Lyon, France

RESEARCH SUMMARY

The first historical *in vivo* study which has revealed the necessity of dietary fats for skin barrier function was conducted in 1929 by Burr and Burr: rats deprived of dietary fats developed typical skin symptoms, such as scaly skin and increased transepidermal water loss. These symptoms reversed after diet supplementation with polyunsatured fatty acids (PUFAs), showing that these essential fatty acids (EFAs) play a major role in skin barrier function and integrity. Numerous *in vivo* studies showed that PUFA intakes improve the skin barrier function in humans, rats or dogs with EFA deficiency or skin diseases. The principal limitation of these studies was to evaluate PUFA supplementation effects on the skin without taking into consideration the differences between the two existing PUFA families, *i.e.*: the omega-6 and the omega-3. However, omega-6 and omega-3 PUFAs play totally distinct roles in many organs, including the skin. The omega-6 family is the most commonly encountered PUFA family in human epidermis. In particular, linoleic acid (LA) is a structural precursor for specific skin elements, known as ceramides, and thus plays the most important role in establishing structure and permeability of the epidermis. On the opposite, alpha-linolenic acid (ALA) and its derivatives of the omega-3 family are not essential compounds of the epidermis but are best known for their immunomodulative properties. Consequently, it is particularly important to distinguish between these two lipid family effects of on the skin barrier function.

A recent study conducted by our research team has demonstrated that the skin barrier function was highly improved in patients receiving intravenous lipid emulsions, which contained only omega-6 PUFAs, compared to patients without intravenous lipid intake. Interestingly, the skin barrier function was not significantly improved in patients receiving intravenous lipid emulsions, which contained both omega-6 and oemga-3 PUFAs. Our results underline the importance of exploring the influence of omega-6 and omega-3 PUFAs on the skin barrier function. Numerous hypotheses can be proposed to explain the observed differences but further investigations still need to be conducted in this field. Establishing of the epidermal barrier and improvement of its function is a priority during development and in individuals requiring skin repair, *e.g.*, premature neonates or severely burn patients. Investigations on PUFA supplementation should further improve our knowledge on mechanisms of omega-6 and omega-3 PUFAs incorporation into skin, leading to the improvement of the epidermal barrier function. New hypotheses could then be proposed for comprehension and treatment of widespread dermatological diseases.

In: Biochemistry Research Summaries. Volume 1
Editors: James T. Walker and Catherine M. Jones
ISBN: 978-1-53610-895-8

Chapter 154

PROTECTIVE EFFECT OF OMEGA-3 POLYUNSATURATED FATTY ACIDS AGAINST ANTIBIOTIC, ANTIHYPERTENSIVE AND ANTI-CANCER DRUG-INDUCED NEPHROTOXICITY AND OXIDATIVE DAMAGE: A FOCUS ON CISPLATIN[#]

Ashreeb Naqshbandi[1], Md. Wasim Khan[1,2], Sana Rizwan[1], Sara Anees Khan[1], Shuhbha Priyamvada[1] and Ahad Noor Khan Yusufi[1] *

[1]Department of Biochemistry, Faculty of Life Sciences, Aligarh Muslim University, Aligarh, UP, India
[2]DST INSPIRE Faculty, Cell Biology and Physiology Division, CSIR-Indian Institute of Chemical Biology, Kolkata, India

RESEARCH SUMMARY

Long-term clinical use of gentamicin (GM), sodium nitroprusside (SNP) and cisplatin (CP) is limited by the development of nephrotoxicity and oxidative damage. Dietary consumption of omega-3-polyunsaturated fatty acids (PUFA) and/or fish oil (FO) is known to have extensive health benefits. They have been shown to retard the progression of various forms of cancers, depression, diabetes, arthritis, and asthma and cardiovascular, neurological, and renal disorders and have been found to improve kidney functions. Here we discussed the protective effects of FO enriched in ω-3 PUFA consumption on GM-, SNP-, and CP-induced nephrotoxicity and oxidative damage. GM/SNP/CP administration resulted in severe renal damage as evident by an increase in serum creatinine, cholesterol, and blood urea nitrogen.

This chapter was previously published in *Omega-3 Fatty Acids: Chemistry, Dietary Sources and Health Effects*, edited by Md. Wasim Khan, New York: Nova Publishers Inc., 2013.

* Corresponding Author: Professor A. N. K. Yusufi, (Retired). Department of Biochemistry, Faculty of Life Sciences. Aligarh Muslim University, Aligarh-202002, UP, India. E-mail: yusufi@lycos.com.

Also, GM/SNP/CP administration led to a decrease in certain enzymes of carbohydrate metabolism, brush border membrane (BBM), and oxidative stress. BBM Pi transport in rat kidney was also reduced. However, dietary FO consumption markedly reversed GM/SNP/CP-induced adverse alterations by increasing energy metabolism, BBM integrity, and Pi transport. These drugs are known to generate ROS/RNOS that cause damage to mitochondria, microsomes, peroxisomes, lysosomes and BBM of renal proximal tubules thus alter their functions. Dietary ω-3 fatty acids affect membrane organization, fluidity, permeability and other properties by altering membrane fatty acids composition. FO-induced changes, at least in part, can also be attributed to improved antioxidant defense mechanisms that accelerate repair and regeneration of damaged cellular membranes. In conclusion, we propose that dietary FO consumption may provide a cushion for prolonged clinical use of these drugs by ameliorating drug-induced nephrotoxic and other deleterious effects.

In: Biochemistry Research Summaries. Volume 1
Editors: James T. Walker and Catherine M. Jones
ISBN: 978-1-53610-895-8

Chapter 155

ROLE OF SUPPLEMENTATION WITH OMEGA-3 FATTY ACIDS IN PREVENTING CARDIAC ARRHYTHMIAS AND SUDDEN CARDIAC DEATH[#]

***Bernhard Rauch*[*,1] and Jochen Senges*[2]**
[1]ZAR- Zentrum für Ambulante Rehabilitation am Klinikum der Stadt Ludwigshafen
[2]Institut für Herzinfarktforschung Ludwigshafen, Germany

RESEARCH SUMMARY

For several decades supplementation with omega-3 polyunsaturated fatty acids (Ω-3) has been believed to reduce cardiovascular risk in primary and secondary prevention, and potential risk reduction especially was attributed to prevention of tachyarrhythmias and sudden cardiac death (SCD). However, recent randomized controlled trials and meta-analyses challenge a general efficacy of Ω-3 supplementation in reducing cardiovascular risk. Especially, an anti-arrhythmic effect as proposed in earlier studies cannot be expected in general, and under certain conditions even pro-arrhythmic effects may occur. Based on molecular interactions of Ω-3 at the cellular level and results of earlier and recent experimental animal studies the following chapter will critically discuss recent clinical studies, emphasizing the individual background conditions that may determine the clinical outcome of Ω-3 supplementation. By this way the hypothesis will be delineated, that the efficacy of Ω-3 supplementation to prevent cardiac arrhythmias strongly depends on the underlying clinical and pharmacological conditions and – at the present stage of knowledge - may be difficult to be predicted in some cases.

[#] This chapter was previously published in *Omega-3 Fatty Acids: Chemistry, Dietary Sources and Health Effects*, edited by Md. Wasim Khan, New York: Nova Publishers Inc., 2013.

[*] Corresponding author address: Email: rauch@zar-kardio-ludwigshafen.de.

In: Biochemistry Research Summaries. Volume 1
Editors: James T. Walker and Catherine M. Jones

ISBN: 978-1-53610-895-8

Chapter 156

OMEGA-3 FATTY ACIDS: ROLE IN INTESTINAL ADAPTATION AND RECOVERY[#]

Igor Sukhotnik*
Technion-Israel Institute of Technology, the Ruth and Bruce Rappaport Faculty of Medicine, Department of Pediatric Surgery and Pathology, Bnai Zion Medical Center, Haifa, Israel

RESEARCH SUMMARY

Over the past decades, growing evidence from several clinical and experimental studies indicates that omega-3 polyunsaturated fatty acids (n-3PUFA) can reduce not only the activity of inflammatory processes but may also improve the immune system and protect the organism against the negative effect of the systemic inflammatory response syndrome. Dietary n-3PUFAs also decrease the production of different proinflammatory mediators such as IL-1,IL-2, IL-6 and tumor necrosis factor-alpha. The positive anti-inflammatory effect of n-3PUFAs has been described in several inflammatory conditions such as inflammatory bowel disease, rheumatoid arthritis, and hepatitis. In this report, we discuss the use of a fish oil-based lipid emulsion in the treatment of different gastrointestinal disorders including parenteral nutrition-associated liver disease, short bowel syndrome, intestinal ischemia-reperfusion, chemotherapy-induced mucositis, microvillous inclusion disease etc.

The mechanisms of action of n-3PUFAs on intestinal cell turnover (proliferation and apoptosis) as well as the effects of this agent on intestinal re-growth after bowel resection and intestinal recovery following intestinal ischemia or chemotherapy-induced damage in rodent models are discussed. Such knowledge may allow us to exploit omega-3 polyunsaturated fatty acids that act as gut trophic agent for the bowel mucosa and suggest new therapeutic strategies to maintain gut integrity after massive bowel resection or intestinal injury, eliminate the dependence on total parenteral nutrition, and avoid the need for intestinal transplantation.

[#] This chapter was previously published in *Omega-3 Fatty Acids: Chemistry, Dietary Sources and Health Effects*, edited by Md. Wasim Khan, New York: Nova Publishers Inc., 2013.

* E-mail: igor-dr@internet-zahav.net.

In: Biochemistry Research Summaries. Volume 1
Editors: James T. Walker and Catherine M. Jones
ISBN: 978-1-53610-895-8

Chapter 157

GASTROPROTECTIVE AND ANTINOCICEPTIVE EFFECTS OF OMEGA-3 FATTY ACIDS: POTENTIAL MECHANISMS UNDERLYING THESE EFFECTS[#]

J. M. Jiménez-Andrade[1] and A. E. Chávez-Piña[2,*]
[1]Unidad Académica Multidisciplinaria Reynosa Aztlán,
Universidad Autónoma de Tamaulipas, Reynosa, Tamaulipas, México
[2]Laboratorio de Farmacología, Programa Institucional de Biomedicina Molecular,
Escuela Nacional de Medicina y Homeopatía del Instituto Politécnico Nacional,
México, D. F., México

RESEARCH SUMMARY

Chronic treatment with fish oil has exhibited gastroprotective effects during several insults such as ethanol, aspirin, pyloric ligation, cold-restrain stress and indomethacin. These actions have been attributed to omega-3 long-chain polyunsaturated fatty acids such as docosahexaenoic acid (DHA) and eicosapentaenoic acid (EPA), which are the main compounds present in fish oil. In support with this, DHA has recently exhibited a gastroprotective effect in the gastric damage-induced by indomethacin mediated by a blockade of gastric LTB_4. The gastroprotective properties of EPA have not been explored yet. Additionally other benefits have been attributed to DHA and EPA. Among those effects, antinociceptive effects by DHA and EPA have been reported in different experimental acute pain models. Mechanisms underlying these antinociceptive effects are related to modulation of synthesis of resolvins from the D and E series (agents that stop inflammatory signaling), as well as activation of opioid receptors, blockade of temperature sensitive transient receptors potential ion channels (thermo TRPs), acid sensing ion channels (ASICs) and increased release of endogenous opioids. Likewise, a reduction of proinflammatory mediators (e.g.

[#] This chapter was previously published in *Omega-3 Fatty Acids: Chemistry, Dietary Sources and Health Effects*, edited by Md. Wasim Khan, New York: Nova Publishers Inc., 2013.

[*] Tel: 11(+52)57296000 ext. 55583. E-mail:achavezp@ipn.mx, arapina@yahoo.com.

eicosanoids and cytokines) and/or a change in the cell membrane phospholipid fatty acid composition have been reported to be involved in the analgesic effect of these omega-3 long-chain polyunsaturated fatty acids.

In: Biochemistry Research Summaries. Volume 1
Editors: James T. Walker and Catherine M. Jones
ISBN: 978-1-53610-895-8

Chapter 158

OMEGA-3 DIETARY INTAKE: REVIEW ON SUPPLEMENTATION, HEALTH BENEFITS AND RESOURCES SUSTAINABILITY[#]

Ivone Almeida, Telmo J. R. Fernandes, Bruno M. R. Guimarães and M. Beatriz P. P. Oliveira*
REQUIMTE, Department of Chemical Sciences, Faculty of Pharmacy, University of Porto, Portugal

RESEARCH SUMMARY

Fish, particularly fatty fish, is the primary food source of omega-3 (*n*-3) long-chain polyunsaturated fatty acids (LCPUFA). Due to different circumstances, it is not always possible to ingest adequate amounts of *n*-3 fatty acids (FA) through fish consumption, which has resulted in an increased interest in omega-3 dietary supplements (mainly fish oil supplements). On the potential beneficial health effects of omega-3 supplements, many epidemiologic and clinical studies have been conducted.

These studies have shown that dietary supplementation with *n*-3 PUFA help to prevent a large variety of medical conditions from occurring or progressing, such as cardiovascular disease, cancer depression, maternal/fetal well-being, and Alzheimer's disease. Omega-3 dietary supplements are available in several retail outlets, health stores and pharmacies and provide *n*-3 PUFA mostly as ethyl esters (EE) or as re-esterified triglycerides (rTG) forms. Different formulations are available, with diverse amounts of EPA and DHA, depending on the omega-3 dietary requirements, and based in various recommendations by scientific, governmental and regulatory bodies, regarding *n*-3 PUFA intake.

Since marine resources are the most important readily available and edible source of *n*-3 PUFA, is crucial to understand whether they're being or not over-exploited. Recent data

---

[#] This chapter was previously published in *Omega-3 Fatty Acids: Chemistry, Dietary Sources and Health Effects*, edited by Md. Wasim Khan, New York: Nova Publishers Inc., 2013.

* Corresponding author: beatoliv@ff.up.pt.

suggest that wild-fish catch may be in decline, while the human population is still growing. Aquaculture is probably the most effective way to maintain global consumption of *n*-3 PUFA. One of the major efforts of aquaculture industry is to replace aquafeed marine-derived ingredients for plant sources, since there is a notorious lack of sustain-ability of the wild-harvest resources that composes fishmeal (FM) and fish oil (FO).

In: Biochemistry Research Summaries. Volume 1
Editors: James T. Walker and Catherine M. Jones
ISBN: 978-1-53610-895-8

Chapter 159

QUALITY AND SHELF STABILITY OF OMEGA-3-FAT-SUPPLEMENTED FOODS[#]

Ann H. Barrett*[*], *Danielle J. Anderson, Tshinanne V. Ndou, Katherine Kensil and Karen R. Conca
Performance Optimization Research Team, Combat Feeding Directorate,
US Army Natick Soldier Research, Development and Engineering Center,
Natick, MA, US

RESEARCH SUMMARY

Omega-3 fats are nutritionally important but lacking in the American diet and, particularly, in the diet consumed by military personnel. These fats provide important physiological benefits including reduction of inflammation and improved psychological health, both of significant importance to deployed Warfighters. However, since omega-3 fatty acids are highly prone to oxidation, incorporating them into foods with a necessarily prolonged shelf-life can be problematic; fortification of military ration components, generally required to have a 2-to-3 year shelf-life, is a particular challenge. Thus, our objectives were to evaluate the sensory quality and loss of omega-3 fatty acids through-out accelerated storage of various supplemented shelf-stable foods representing different chemical and packaging environments. Representative low-moisture products (performance bars, packaged w/o oxygen scavengers), intermediate-moisture products (cakes, packaged with oxygen scavengers), and high-moisture (retorted) products were produced each with two levels of encapsulated omega-3 fat, up to 400 mg/serving, and stored for 4 weeks at 49°C and 3 and 6 months at 37°C. Sensory quality, physical properties, retained omega-3 DHA and EPA, and fluorescent volatile (i.e., Schiff base) production were measured. Sensory quality of the cakes and retorted beef lo mein did not diminish throughout storage regardless of omega-3 content. These items also showed negligible loss of omega-3 fat. Performance bars, however, showed

---

[#] This chapter was previously published in *Omega-3 Fatty Acids: Chemistry, Dietary Sources and Health Effects*, edited by Md. Wasim Khan, New York: Nova Publishers Inc., 2013.

[*] Corresponding author: ann.h.barrett.civ@mail.mil.

a 30% decrease in overall quality over 6 months and a 17% loss in omega-3 fat due to oxidation.

In conclusion, extended shelf-life foods supplemented with high, physiologically significant levels of omega-3 fats can be successfully developed, with the provision that oxygen is limited in the carrier products.

In: Biochemistry Research Summaries. Volume 1
Editors: James T. Walker and Catherine M. Jones
ISBN: 978-1-53610-895-8

Chapter 160

OMEGA 3 FATTY ACIDS AND GPR120[#]

N. A. Arivarasu[1], Shubha Priyamvada[1] and Medha Priyadarshini[2,*]
[1]Department of Medicine, University of Illinois, Chicago, US
[2]Department of Biochemistry, F/O Life Sciences, AMU, Aligarh, U.P., India

RESEARCH SUMMARY

Omega-3 fatty acids have various health benefits of which the underlying mechanisms remain unknown. Free fatty acids have been demonstrated to serve as ligands for G protein coupled receptors. GPR120 acts as the receptor for omega-3 fatty acids and mediates most of their physiological effects. Importantly, GPR120 signaling is at the focal point of inflammatory and insulin sensitization pathways, presenting it as a potential therapeutic target. This chapter highlights GPR120 signaling mechanisms with the main emphasis on the physiological effects exerted upon its sensitization with omega-3 fatty acids.

[#] This chapter was previously published in *Omega-3 Fatty Acids: Chemistry, Dietary Sources and Health Effects*, edited by Md. Wasim Khan, New York: Nova Publishers Inc., 2013.
[*] E-mail: medha.priyadarshini@gmail.com.